International Zinc and Zinc Oxide Conferences 2019

Scottsdale, Arizona, USA
24-27 February 2019

ISBN: 978-1-5108-8559-2

Printed from e-media with permission by:

Curran Associates, Inc.
57 Morehouse Lane
Red Hook, NY 12571

Some format issues inherent in the e-media version may also appear in this print version.

Printed with permission by Curran Associates, Inc. (2019)

For permission requests, please contact International Zinc Association
at the address below.

International Zinc Association
2530 Meridian Parkway
Suite 115
Durham, NC 27713
USA

Phone: +1 919 361 4647

contact@zinc.org

Additional copies of this publication are available from:

Curran Associates, Inc.
57 Morehouse Lane
Red Hook, NY 12571 USA
Phone: 845-758-0400
Fax: 845-758-2633
Email: curran@proceedings.com
Web: www.proceedings.com

TABLE OF CONTENTS

2019
INTERNATIONAL
ZINC & ZINC OXIDE
CONFERENCES

FEBRUARY 24-27

The Westin Kierland Resort
Scottsdale, Arizona

International
Zinc Association
Zinc...essential for modern life

DOWNLOAD AND USE THE MOBILE EVENT APP

STEP 1

DOWNLOAD THE APP

Go to the App store of your smartphone, search for the following APP URL https://crowd.cc/s/2qDtq and download the App for free on your smartphone or tablet.

STEP 2

VERIFY YOUR ACCOUNT

You will be prompted to verify your account by confirming your email address. A verification code will be sent to you immediately which you need to enter to access the App.

Allow for push notifications so you can receive instant updates.

STEP 3

BEGIN USING THE APP

Access the main menu to:

- Create your schedule
- View list of attendees & speakers
- View agenda
- Send private messages
- Add your picture and bio

ACCESS THE APP VIA A WEB BROWSER FROM YOUR COMPUTER
and download the presentations after the conference

https://crowd.cc/zincconference2019

SPONSORS

GLENCORE

PEÑOLES

ocean PARTNERS

Pacorini

TRAXYS

Richker Metals Inc.

SAVE THE DATE

for IZA's 2020 International Zinc & Zinc Oxide Conferences

February 16-19, 2020

The Westin Kierland Resort & Spa

Scottsdale, Arizona

SATURDAY, FEBRUARY 23

3:00 - 5:00 PM
Registration Desk Open - Visionaries Hall

SUNDAY, FEBRUARY 24

3:00 - 5:00 PM
Registration Desk Open - Visionaries Hall

6:00 - 7:30 PM
WELCOME RECEPTION - Marshall's Pavillion

MONDAY, FEBRUARY 25

7:00 - 8:30 AM
BREAKFAST - North Foyer, Northern Sky Terrace
additional seating: Greenway, Herberger 1,2

7:30 AM - 12:00 PM
Registration Desk Open - Herberger Foyer West

8:45 - 11:30 AM
GENERAL SESSION:
International Zinc Conference
Herberger 3,4 and 5AB

Introduction
Stephen Wilkinson,
International Zinc Association

KEYNOTE SPEECH
The Impact of Electric Vehicles on the Zinc and Zinc Oxide Industries
Ted Graham, General Motors

Century Mine Re-Start
Patrick Walta, New Century Resources

Zinc Batteries : A bright Future
Ramkumar Krishnan, NantEnergy

TUESDAY, FEBRUARY 26

7:00 - 8:30 AM
BREAKFAST - North Foyer, Northern Sky Terrace
additional seating: Greenway, Herberger 1,2

8:00 AM - 4:00 PM
Registration Desk Open - Herberger Foyer West

8:45 AM - 1:30 PM
NETWORKING TIME
Herberger rooms will be available.

1:30 - 4:00 PM
JOINT SESSION:
Int'l Zinc & Zinc Oxide Industry Conferences
Herberger 4 and 5AB

ZINC MARKET OUTLOOK
Jonathan Leng, Wood MacKenzie
Vivienne Lloyd, Macquarie
Ryan Cochrane, CRU
Duncan Hobbs,Concord Resources
Cindy Wei, BGRIMM Lilan Consulting

6:00 - 7:30 PM
RECEPTION - Northern Sky Terrace

NOTE:
Presentations will be made available on the App after the conference

BADGES ARE REQUIRED AT ALL TIMES!

WEDNESDAY, FEBRUARY 27

8:00 AM - 3:00 PM
Registration Desk Open - Herberger Foyer West

9:00 AM - 12:00 PM
GENERAL SESSION:
Int'l Zinc Oxide Industry Conference
Herberger 4 and 5AB

Zinc Market Overview
Heike Held, IHS Markit

Zinc Oxide in Fertilizers
Taylor Pursell, Pursell Agri-Tech

Animal Feed/Supplementation:
Market Overview
Russell Wyllie, Zinpro Corporation

10:30 - 11:00 AM
BREAK - Herberger Foyer

Zinc Oxide Market:
Opportunities and Challenges
Mohit Sharma, Zochem LLC

Regulatory Update:
Challenges to the Market
Eric Van Genderen,
International Zinc Association

12:00 PM - 1:30 PM
LUNCH - Northern Sky Terrace

2:00 - 4:00 PM
Zinc Oxide Committee Meeting - Greenway

international
zinc association

2019

International ZINC & ZINC OXIDE Conferences

February 24-27

Westin Kierland Resort
Scottsdale, Arizona, USA

international zinc association

International Zinc Conference – 27 Years Strong

George Vary – Executive Director – AZA

George organized the first International Zinc Conference in 1992, and served as its colorful MC until his retirement in 2012.

Sadly, George passed away last summer, leaving an indelible impression on this conference, the zinc industry, and all who had the pleasure of knowing him.

George Folk Vary
1949 - 2018

Statement of Compliance

1. **Meeting of the International Zinc Association, Monday 25 February 2019.**

2. The purpose of the meeting was to address issues concerning zinc producers; to promote the public awareness of and interest in the zinc industry; and to monitor governmental activities and related tasks.

3. No discussions were held, formally or informally, during specified meeting times or otherwise, involving, directly or indirectly, express or implicit agreements or understandings related to: (a) any company's price; (b) any company's terms or conditions of sale; (c) any company's production or sales levels; (d) any company's wages or salaries; (e) the division or allocation of customers or geographic markets; or (f) customer or suppliers boycotts.

4. The International Zinc Association as a group makes no recommendations of any kind and has not reached any agreements or understandings with respect to an individual company's prices, terms or conditions of sale, production or sales levels, wages, salaries, customers or suppliers.

5. I have read the foregoing and it is true to the best of my knowledge and belief. This statement is freely given without threat or reprisal of promise of benefit.

Name	Title
Company	Date

Sustainability Resources

Galvanized
Autobodies

Thermal
Spraying
ZINC THERMAL SPRAYING:
CORROSION PROTECTION OF STEEL AT SEA
• Proven long-term corrosion protection of steel in marine environments
• Exteremely abrasion resistant
• Application in shop or field
• Superb adhesion to steel
• Excellent base for paint
• Instant drying
• Significant life-cycle cost savings through minimal maintenance
• Duplex coating with paint provides even more protection
• There are ISO and ASTM standards for thermal spraying (ISO 2063- 1, ISO 2063-2 and ASTM B833-13)

Galvanized
Rebar

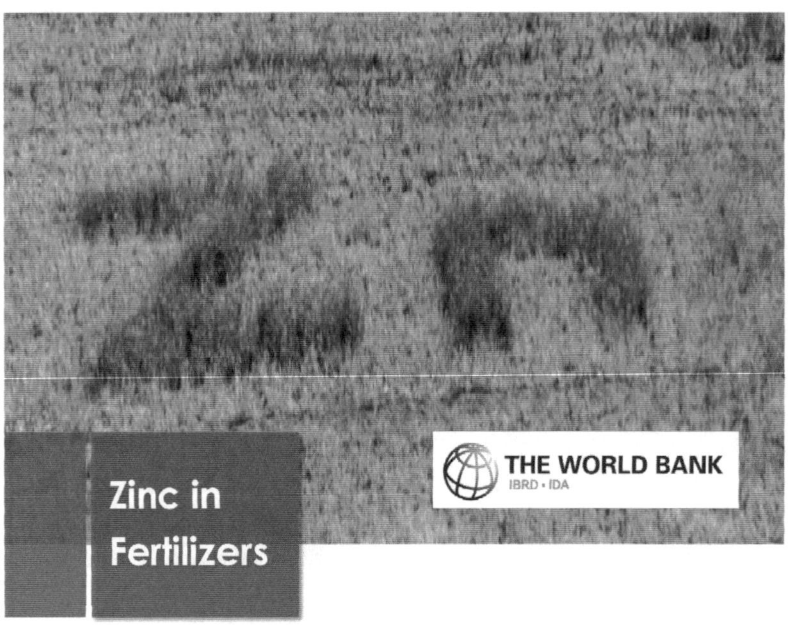
Zinc in
Fertilizers
THE WORLD BANK
IBRD • IDA

New Programs

SAVE THE DATE

IZA's 3rd annual

ZINC CONFERENCE – EUROPE

15-17 May, 2019 - Plovdiv, Bulgaria

plovdiv
together
2019
EUROPEAN CAPITAL
OF CULTURE

International
Zinc Association
Zinc...essential for modern life

Zinc… essential for modern life

北矿力澜
BLC BGRIMM LILAN CONSULTING

China's Zinc Market

Cindy Wei

BGRIMM LILAN CONSULTING

Feb 2019

Major concerns of Chinese zinc industry

- **How will newly EP related regulations impact China zinc industry?**
  - Green mine construction
  - Tl content up-limits in discharged wastewater
  - Imports ban of metal slag/residue/dust etc
  - New standards of impurity up-limits in the Zn concentrate
- **How will China zinc mine production grow?**
- **What are the main factors restricting Chinese zinc production?**
  - Special emission up-limits
  - Construction of leaching residue (HW) disposal line

EP!
EP!!
EP!!!

BGRIMM Lilan Consulting

- **BGRIMM -** Beijing General Research Institute of Mining & Metallurgy
  - a state-owned company
  - founded in 1956
  - a products & technology supplier in the areas of mining, mineral processing, non-ferrous metallurgy, industrial explosives & blasting, processing equipments and reagents, magnet materials.
  - a service provider for resource evaluation, technical audits and diagnostics, Due diligence, process development, Environment Impact Assessment (EIA), feasibility study, engineering design and EP/ EPCM /EPC etc.
- **BLC**
  - a share-holding company of BGRIMM
  - an internationally renowned metals research consultancy to provide marketing intelligences, cost analysis, consulting services on China's base metal industry.
  - Depend on a countrywide network of mines, smelters, refineries, fabricators, downstream end users and industry associations
  - To give clients the insight of China's base metal industry.
  - to provide independent and professional in-depth advices and consulting services on assets, companies and markets.

Zinc mine and smelter production

China zinc mine production

Production forecast revision footprint in 2017-2018 Unit: Kt-Zn

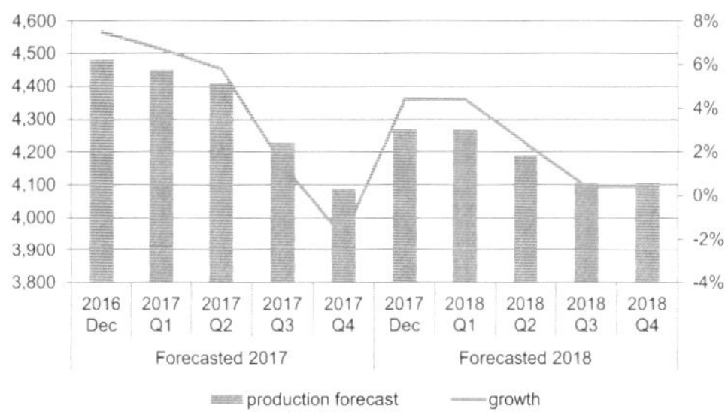

2018 0.4% YoY (+16kt-Zn) to 4,106kt

- All mines in Hunan Huayuan were shut down (-50kt-Zn at least)
- Intensive EP inspections in May & October in Sichuan, Inner Mongolia and Yunnan etc
- Falling ore grades at some big mines (Zijin's Sanguikou, Dongshengmiao and Lanping Pb-Zn mine, etc.)
- Lower than expected production at the newly commissioned mining projects
  - ➢ 6 mining projects (69kt/a-Zn capacity) were commissioned with only 24.7kt-Zn produced

5 BLC

Outlook of China zinc mine production

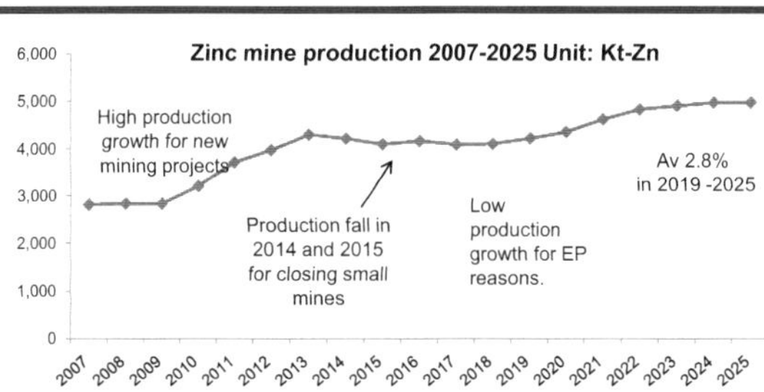

Source: BGRIMM Lilan consulting (BLC)

New zinc mining projects 2017-2022 Unit: Kt-Zn

Source: BGRIMM Lilan consulting (BLC)

<u>2019 + 2.8% YoY (+114kt-Zn) to 4,220kt</u>

- 33.6kt-Zn production increases from new mines (69 kt/a-Zn) in 2018
- 75kt-Zn production from 11 new projects (267 kt/a-Zn) in 2019
  - ➢ Guojin Mining's Bachagou West mining project (24.9kt/a-Zn) in Inner Mongolia
  - ➢ Cuihongshan Mining's 700ktpa mining/milling project in Heilongjiang
- Mines in Hunan Huayuan production resume (not so optimistic)
- Probable less mine interruptions by EP and safety inspections ???
- Unexpected interruptions-Yinman's accident on 23rd Feb

6 BLC

Green mine construction-The 1st national standard for green mine construction issued on Oct., 2018

- to promote the **rational use of resources**, energy conservation and emissions reduction, the protection of the ecological environment and the harmony of the mineral land
- The greening of the mining area shall be harmonious with the surrounding natural environment and landscape. The greening plants shall reasonably match, and the greening coverage rate of the mining area shall reach 100%.
- The closed tailings dams and waste dumps from open-pit mining shall be reclaimed and "greened". Isolated green belts shall be set on both sides of the road in the mining area in line with local conditions.

The lowest mining recover rate for green-standard Pb&Zn mine

Mining Method	Recovery rate
Open pit mining	
Large Mine	95%
Small and medium-sized mines or mines with varied ore body shapes, thin ore bodies and instable ore rocks	92%

Underground mining	Pb&Zn Sulfide ores			Pb&Zn Sulfide bulk ores			Pb&Zn oxide ores		
Thickness of the ore body	Pb+Zn≥9.0%	Pb+Zn 4.5%~9.0%	Pb+Zn≤0.45%	Pb+Zn≥11.5%	Pb+Zn 6.0%~11.5%	Pb+Zn≤6.0%	Pb+Zn≥14%	Pb+Zn 7.5%~14.0%	Pb+Zn≤7.5%
≤5m	88%	80%	75%	88%	80%	75%	88%	80%	75%
5~15m	92%	83%	80%	92%	83%	80%	92%	83%	80%
≥15m	92%	85%	85%	92%	85%	85%	92%	85%	85%

7 BLC

Zinc Concentrate Imports

Zinc Concentrate Imports 2016-2018

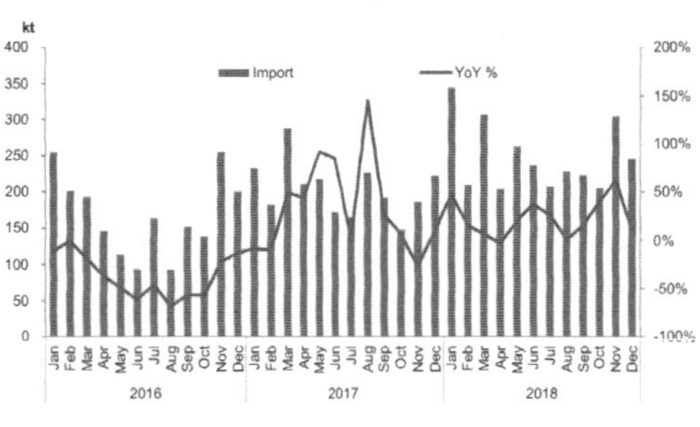

Source: BGRIMM Lilan Consulting

Prices of domestic conc. and imported conc. 2015-2019

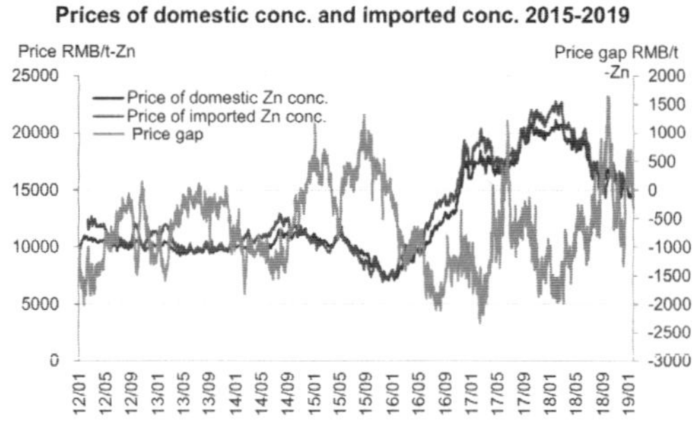

Source: BGRIMM Lilan Consulting

2018 Concentrate imports +22% YoY (533kt) to 2.97Mt

- Increased TCs encouraged some smelter to use more imported concentrate (eg.Zhuzhou, Zhongjin Lingnan)
- Production increase of Chinese invested overseas zinc mine supported high concentrate imports
- We saw increased Zn concentrate inventories in ports (over 200kt)

8 BLC

Chinese invested overseas zinc mine

- Total 18 Chinese invested overseas zinc mines
- 9 operations: 6 in Asia; 3 in Australia
- Chinese invested overseas zinc mine production in 2018 increased to over 600kt-Zn from 350kt-Zn in 2016

- All of these projects were invested by Chinese large-size SOEs or publicly list companies. CNFC or its branch NFC is the largest investor for Zn mine assets overseas.
- Four notable overseas Zn-related asset acquisitions by Chinese companies were completed in 2018.
  - Zjin's US$1.41 billion acquisition deal to purchase all of Nevsun Resources issued and outstanding shares.
  - CITIC Metal's US$560 million acquisition deal for a 19.9% stake in Ivanhoe Mines, which owns Kipushi project in the DRC
  - China Non-Ferrous Metals (NFC)'s US$ 198 million acquisition deal for a 51% stake in PT Dairi Prima Mineral (DPM) in Indonesia.
  - Zhongjin Lingnan's Perilya Limited A$14.9 million acquisition deal for a 19.9% stake in Myanmar Metals Limited (MYL).

Imports ban of metal slag/residue/dust etc

China has banned Zn slag residues imports (Since 31 Dec 2017)

HS Code	Item
2620110000	Hard Zn slag, ash and residue
2620190000	Other Zn slag, dust, ash and residue

BLUE SKY Actions

防城海关退运禁止进口固体废物1556.12吨（图）

In Jan, 2019, 1.5kt imported Zn residue, which was claimed as ZnO, was rejected by the Fangchenggang customs

连云港海关查获2009吨进口冶炼废弃物（图）

In Sep, 2018, Lianyungang customs seized 2009 t smelting slag with 76 containers, which were claimed as crude ZnO. It was detected as Arc furnace ZnO dust.

The imports of Pb& Zn concentrate

- Imports of Pb and Zn concentrate also follow "The specification for limit on harmful element content of the heavy metal concentrates" **(GB20424-2006)** , was implemented as a compulsory standard between 2006 to Oct., 2016.
- China Non-ferrous Industry Association (CNIA) is drafting a "specification for limit on harmful element content of the Zn concentrate" (**2018-040-T/CNIA**). This will be a recommended standard, not a compulsory standard. The new standard, scheduled to be circulated in 2019, will only apply to Zn concentrate.
- CNIA has send questionnaires to major Zn smelters for their comments on impurity up-limits in the Zn concentrate. Besides the three major impurities (As, Cd and Hg), other impurities (**Tl, F and Cl**) have been also listed in the questionnaires.
- **It is highly possible that Tl in the imported Zn concentrate would be restricted by the new standard. We heard that imported high Tl content Zn concentrate has been rejected by a Chinese smelter in Guangudong.**

The up-limits for the imported Zn concentrate and Bulk concentrate (GB 20424-2006)

	Impurities %		
Concentrate	As	Cd	Hg
Zn concentrate	0.6	0.3	0.06
Pb Concentrate	0.7		0.05
Bulk concentrate (Pb&Zn)	0.45	0.4	0.05

BLC

Refined Zn production

Zinc smelter production 2007-2022 Unit: Kt

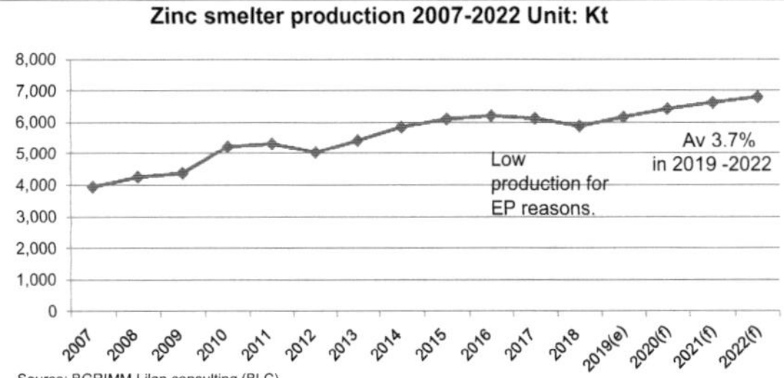

Source: BGRIMM Lilan consulting (BLC)

New zinc smelter projects 2015~2021

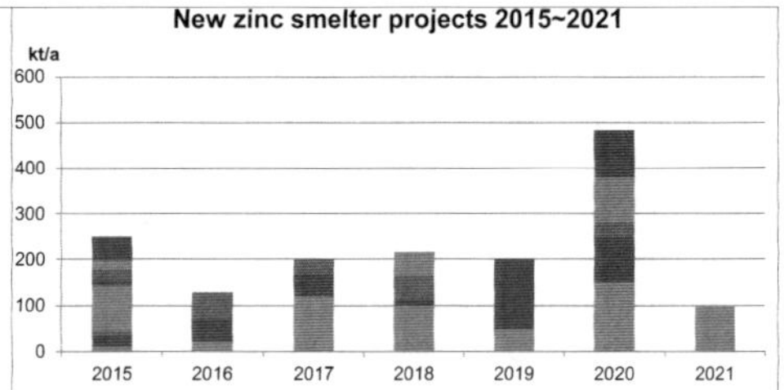

Source: BGRIMM Lilan consulting (BLC)

2018 -3.9% YoY (241kt) to 5,878kt

- Zhuzhou smelter reduced production rate since mid-year, prepare for relocation
- Three Zn smelters in Huanyuan Hunan were shutdown for EP issues
- Smelter maintenance shutdowns in Q2 due to low TCs
- HW disposal issues (eg. Danxia, Luoping Zinc & Electricity and Hanzhong Zinc etc.)
- EP inspections in Fen-wei 11 cities (Dongling suspended one VR line)
- Secondary zinc production was down by 1.8% YoY in 2018 (eg. Xiangyun Feilong)
- 4 new projects - longer-than-expected commissioning, low production contribution

2019 +4.8% YoY (285kt) to 6,163kt

- Strong production rate at most smelters in Q1 for the high TCs should be offset by Zhuzhou smelter's low production.
- Current high TCs should be at least till to the mid-2019. Capacity ramp-up at Zhuzhou as well as increased production rate at most smelters for high TCs might drive Zn production to increase in Q2.
- Most HW disposal projects will be completed in Q3, supporting higher production in Q4.
- Low production contribution from two new projects (200 kt/a-Zn)

BLC

HW issue on Zn production

Breakdown of China's Zn production process

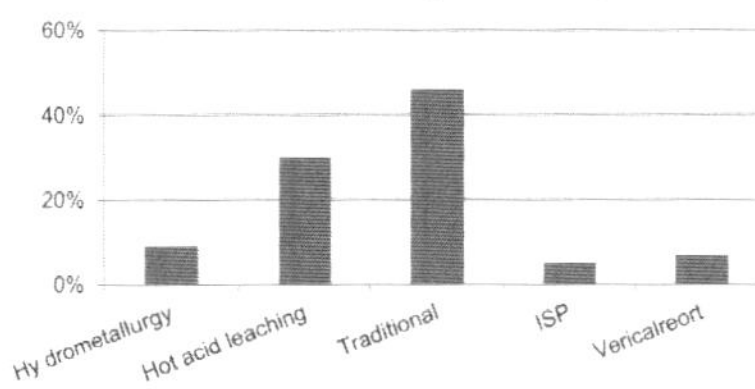

<u>HW from Zn production (HW-48) -2016 version</u>

- 14 kinds of wastes generates from zinc smelters
- Almost all type of solid wastes from zinc hydrometallurgy process are HW due to the easily dissoluble characters of lead and silver contained in the residues
  - Leaching residue from the conventional leaching;
  - Jarosite residue from hot acid leaching;
- Zn ingot casting dross;
- Waste residue from the extraction of other value metals in lead-zinc smelters;
- Dust collected by dust collector in lead-zinc smelting process;

- In June, CCTV reported that Luoping stockpiled 100kt Pb residue without disposal. Luoping's 100kt/a Zn smelter was shut down in Jun-Sep 2018.
- Hanzhong ran at 30-40% of nameplate capacity because of bottle neck of leaching residue disposal.
- Six HW disposal lines have been in the pipe line, covering 1,310kt Zn smelting capacity. Normally, construction period of a HW disposal line is **about 9~12 months.**
- Investment of such leaching residue disposal line is **at least 50 million RMB.**

Tl Pollution & Policy

- Thallium is normally combined with copper, lead, zinc, and other sulfide ores.
- Since environmental pollution caused by thallium is not as common as As, Cd, Pb, Hg and so on, **thallium was not monitored by environmental protection departments at all levels before 2010.**
- **We saw some Tl pollution accidents by Zn smelters between 2010-2017**
  - **Zhongjin Lingnan's Shaoguan** smelter-- Oct, 2010
  - Hezhou Huiwei Mineral Processing Plant -- Jul, 2013
  - Hanzhong smelter – May, 2017
- **Hunan** published the provincial Emission Standard of Thallium for Industry Wastewater, Jan 1, 2015
- **Guangdong** issued the Emission Standard of Thallium for Industry Wastewater, Oct 1, 2017
- **On Sept 20, 2018, MEE published the revised "Emission standard of pollutants for lead and zinc industry (GB 25466-2010)".**
  - Total Tl content in discharged wastewater should be lower **than 5 µg/L**.
  - The existing zinc and lead smelters are required to execute since Sept 1, 2019
  - New smelters have to follow up once it is implemented.
- **With the new standard, about 570-1000 RMB/t-water CAPEX is needed to install related waste water treatment capacity and OPEX is about 3-9 RMB/t-water per day.**

Special emission up-limits for Zn smelters

Emission standard for Pb/Zn industry

Pollutants	GB 25467-2010		Special emission up-limits 2013
	1997-2010	2010	
SO2	800~960	400	**100**
PM 10	100	80	**10**
Nitric oxide	0	0	**100**
Sulfuric acid mist	45	40	**20**
Pb & Pb compounds	10	8	**8**
Hg & Hg compounds	1	0.05	**0.05**

- By 2020, 15 Zn smelters with 2,363 kt/a smelting capacity will have been involved in implementing the special emission up-limits.
- Most Zn smelters will upgrade its off-gas collecting system to reduce PM and SO2 emission.

time schedule for special emission up-limits Vs. Involved Zinc smelters

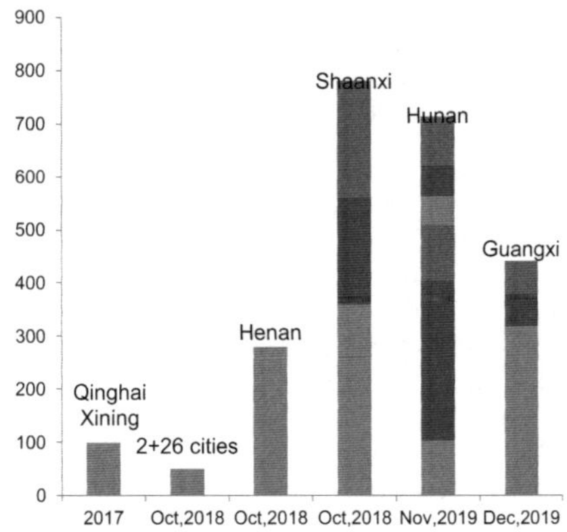

Zinc Demand

Refined Zinc Consumption I

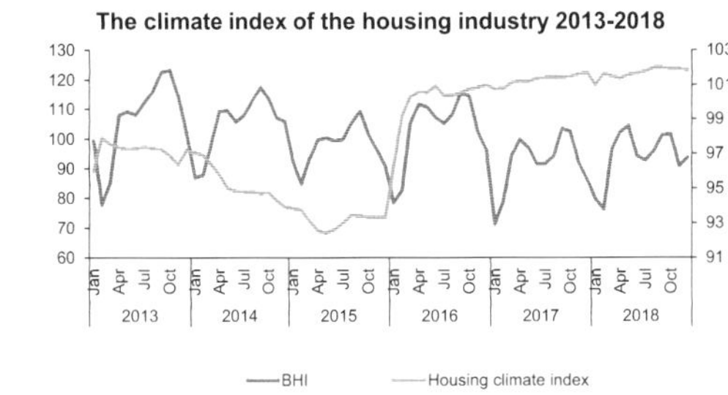

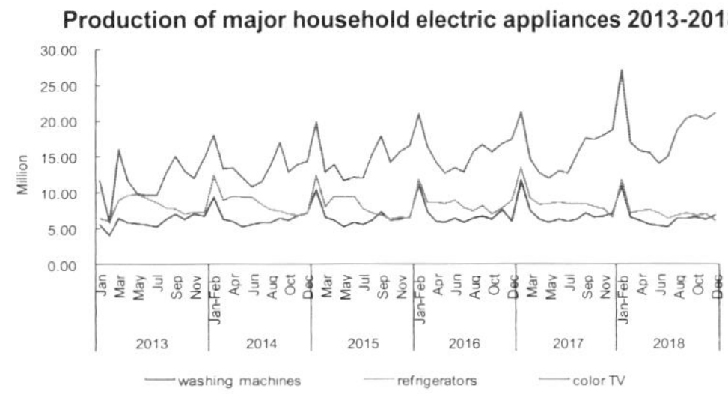

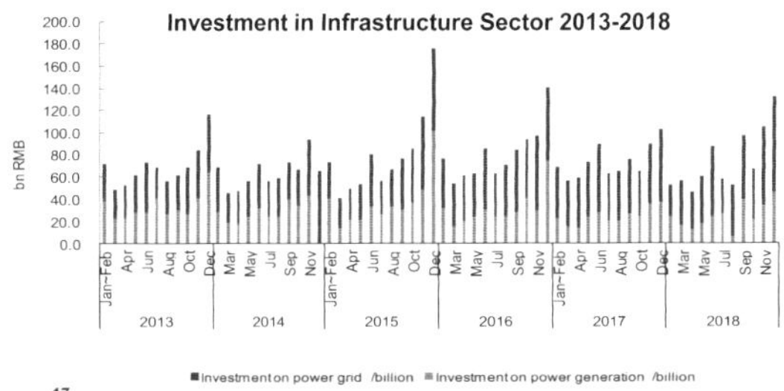

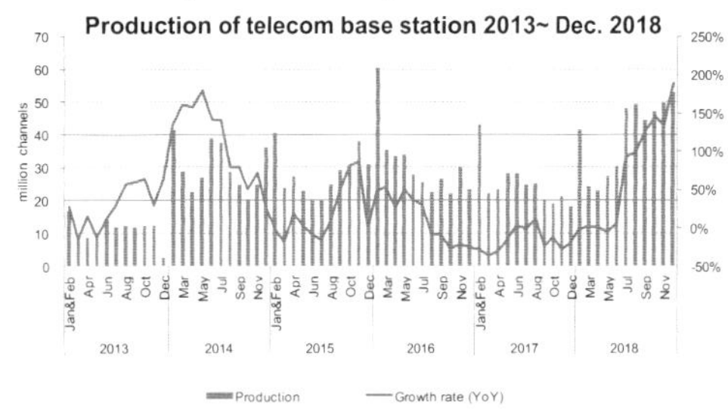

17

Source: NBS

Refined Zinc Consumption II

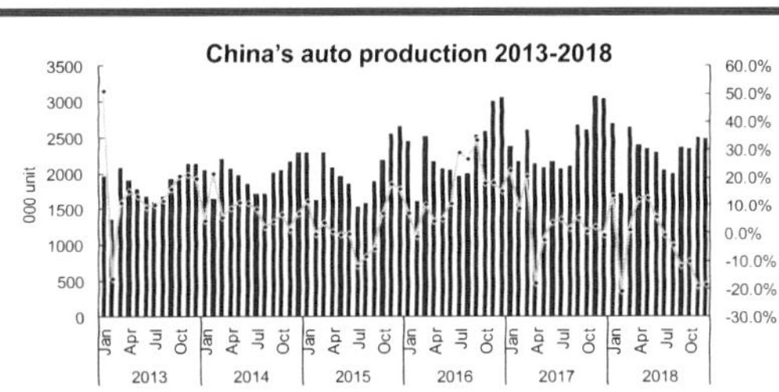

Source: China Association of Automobile Manufactures

Source: NBS

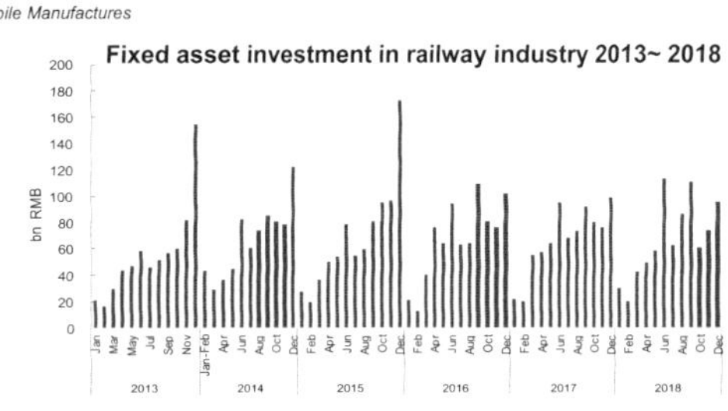

Source: The Ministry of Railways of the People's Republic of China

18

BLC

Conclusion and outlook for 2019

		2017	2018	2019f
Zn mine production		**4,090**	**4,106**	**4,220**
	YoY GR	*-1.9%*	*0.4%*	*2.8%*
Net import of Zn concentrate		1,180	1,458	1,560
	YoY GR	*27%*	*24%*	*7%*
Refined Zn production		**6,119**	**5,878**	**6,163**
	YoY GR	*-1.4%*	*-3.9%*	*4.8%*
Net import of zinc ingot and alloy		767	797	836
	YoY GR	*53%*	*4%*	*5%*
Refined Zn consumption		**6,966**	**6,925**	**6,999**
	YoY GR	*0.5%*	*-0.6%*	*1.1%*

Source: BGRIMM Lilan Consulting

Supply

- Eased domestic concentrate and secondary materials supply
  - More new mining projects
  - Increasing concentrate imports
- Less production interruption in both mines and smelters for EP inspections

Demand

- Improved housing completion
- Improved infrastructure
- Recovered auto production drive by new stimulating policies
- Pressures on exports given appreciation of RMB to be continuing

19 BLC

Brief description

- Zinc Mine Production
- The status of Chinese zinc mining projects overview
- Zinc Smelter Production
- Refined Zinc Consumption

Frequency and delivery

- This product includes four issues with annual.
- PDF delivery to subscribers' inbox on 31 Jan. 30 Apr. 31 Oct. and 31 Jan. in electric saved file in English.

Cost

- **US$7,500** for annual subscription

20

THANK YOU!

Cindy Wei
Email: wei_yf@lilanconsulting.com
Tel: +86-10-63299741
Fax: +86-10-63299744

2018-19 zinc market dynamics in three words – strong, struggling, soft

IZA International Zinc Conference

Scottsdale – 26th February 2019

Concord Resources Limited

Duncan Hobbs
+44 20 3879 3966
duncan.hobbs@concordltd.com

CONCORD

Disclaimer

CONCORD 1

Concord Resources Limited
Who we are

- **Founded in 2015** Concord is a global resources merchant focused on the supply chain, financing and risk management of **non-ferrous metals and associated raw materials**, led by an experienced and entrepreneurial management team
- Concord is a global company, operating from its **headquarters in London** and through its principal **offices in New York and Hong Kong**. Our experienced team work with resource producers and consumers on spot transactions out to multi-year, structured agreements
- Concord's capabilities include **global sourcing, marketing and management of all the required logistics and finance** to optimise our clients' supply chains from long-term partnership marketing arrangements to single container deliveries
- Concord brings substantial cross-commodity, structuring and asset optimisation experience of its founding team to work for the **benefit of its customers and partners**
- Concord is an **independent company**, and combines its founding management's experience with the commodity and financial industry expertise of its shareholders
- For more information please visit our website

CONCORD 2

2018-19 zinc market dynamics in three words – strong, struggling, soft

- 2018 market review
- 2019 market outlook
- Summary & conclusions

CONCORD 3

2018-19 zinc market dynamics in three words – strong, struggling, soft

- 2018 market review
- 2019 market outlook
- Summary & conclusions

CONCORD

4

Strong – zinc price posted highest annual average in 11 years in 2018

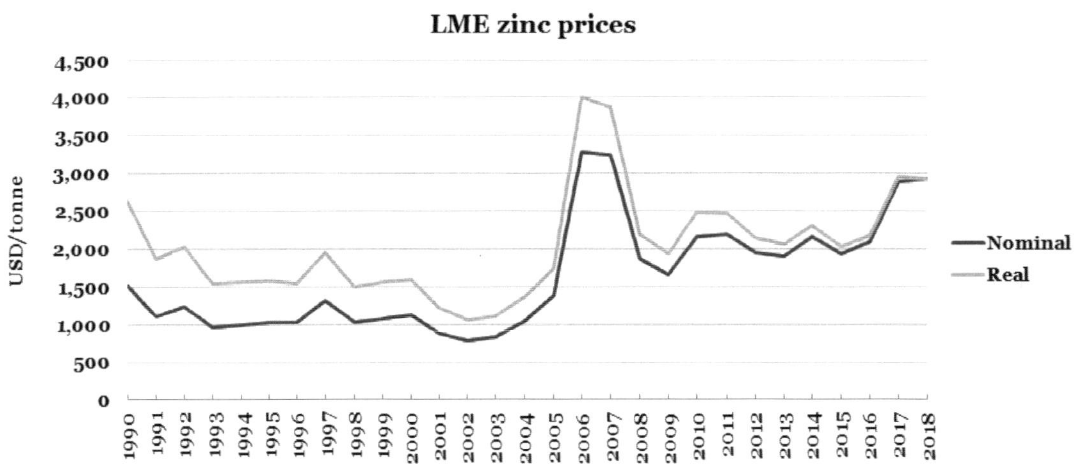

- LME cash zinc price averaged > USD2,900/tonne in 2018, up > 50% vs recent low in 2015
- 2006-2007 records were > 10% higher in nominal terms, > 30% higher in real terms

Note: LME cash settlement prices. Real prices basis 2018 USD, calculated using US GDP deflator
Source: LME, US BEA, Concord Resources Limited, February 2019

CONCORD

5

Despite zinc prices declining with wider market through much of last year

- LME metals prices under pressure from mounting worries over the economic outlook
- Added pressure on zinc from forecast rise in supply; finished year down > 25% vs LMEX ~18%

Note: LME cash settlement prices
Source: LME, Concord Resources Limited, February 2019

CONCORD 6

Struggling – zinc mine output up last year but still below previous peak

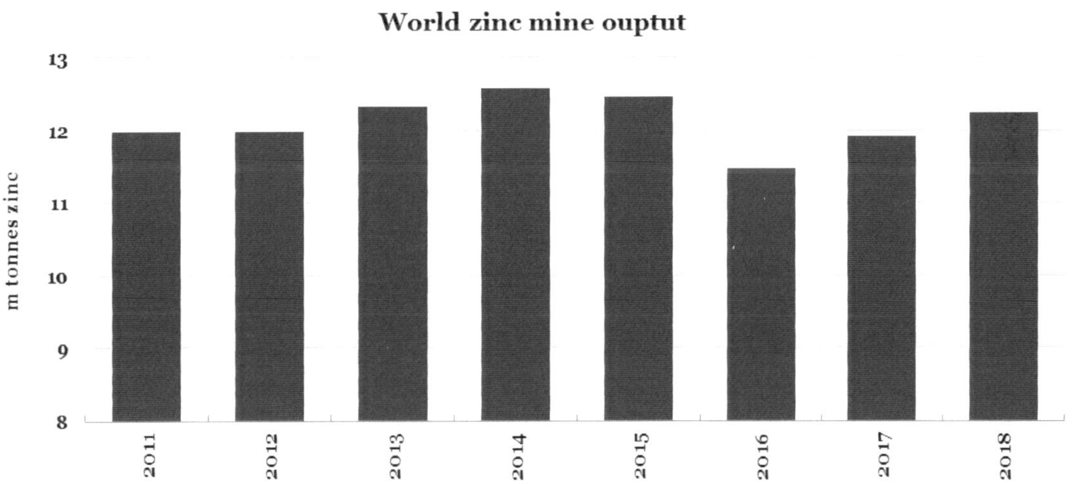

- World zinc mine output up < 3% YoY to ~12.3m tonnes zinc in 2018; new projects, restarts
- However, output still below mid-decade peak of > 12.5mtpa preceding key mine closures

Source: CRU, Concord Resources Limited, February 2019

CONCORD 7

Struggling – zinc smelter production also dropped last year

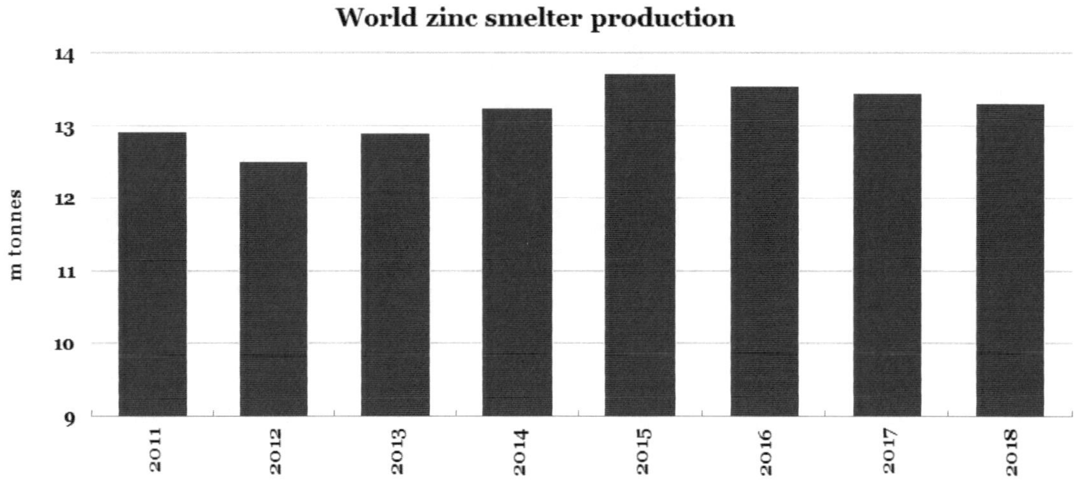

- World zinc metal output fell by ~150,000 tonnes (-1%) YoY to ~13.3m tonnes in 2018
- 3rd consecutive decline, leaving output ~3% below peak of 13.7m tonnes in 2015

Source: CRU, Concord Resources Limited, February 2019

CONCORD

8

World zinc mine and smelter output was shaped by outcomes in China

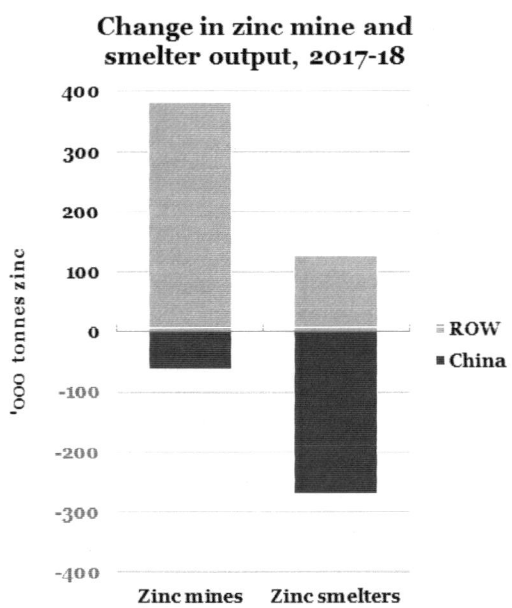

- China is the world's largest zinc miner (34% share in 2018) and smelter (43%)
- China's zinc mine output fell -1.5% in 2018, limiting increase at world level, and has now fallen in 3 / last 4 years
- China's zinc smelter production fell almost 300,000 tonnes in 2018, more than double the increase recorded in rest of world
- Drop in mine and smelter output attributed in part to tighter environmental controls
- Pressure on smelters amplified by low TCs and declining demand

Source: CRU, NBS, Concord Resources Limited, February 2019

CONCORD

9

Recently China's zinc mine output hasn't responded to financial incentive as in the past

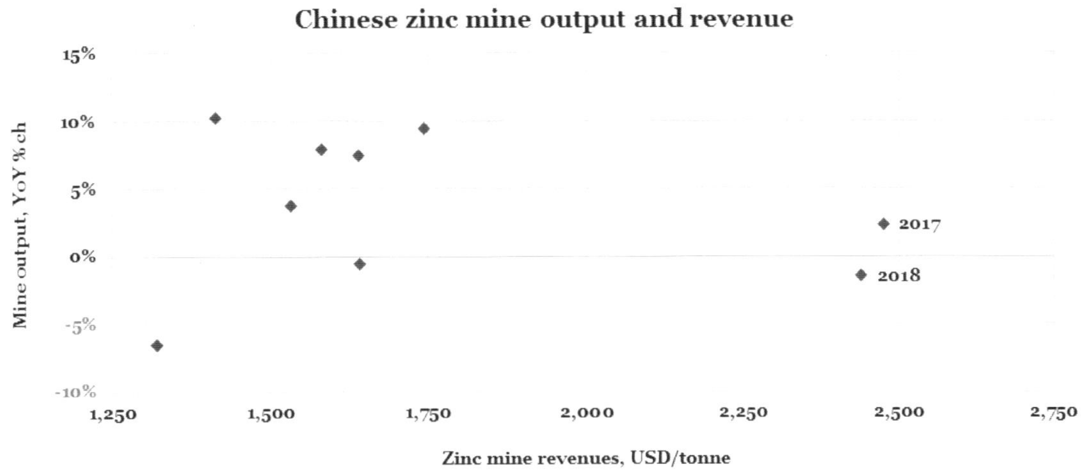

- Higher mine revenues once brought stronger output growth; relationship broke down in 2017-18
- 2010-16: CAGR ~4% @ ave ~USD1,550/tonne zinc; 2017-18: CAGR < 0.5% @ > USD2,450

Source: CRU, FastMarkets, LME, Concord Resources Limited, February 2019

CONCORD

10

Soft – zinc consumption dipped in 2018, 1st fall since GFC in 2009

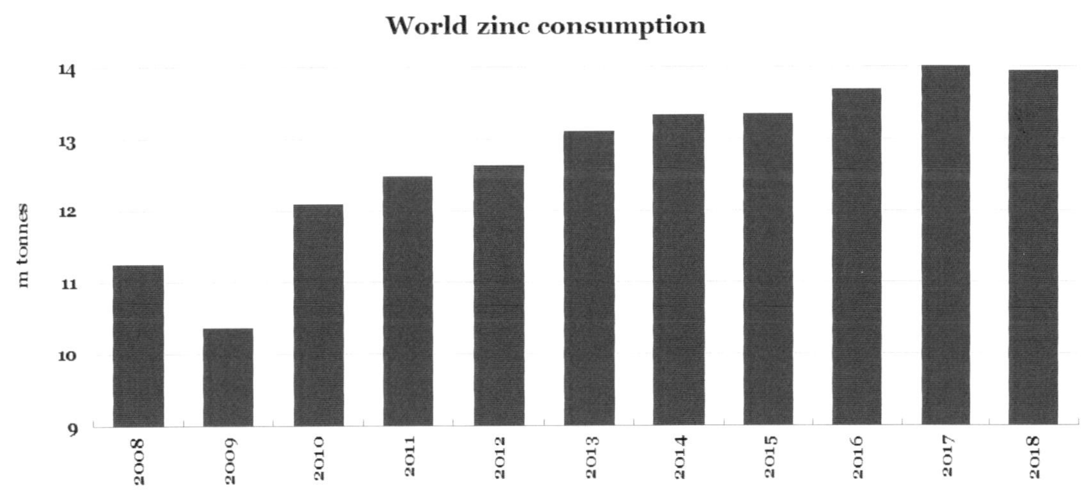

- World total -1% YoY to < 14m tonnes; China -1.5%; Europe -0.5%; USA about flat
- Slowing world auto sales / output; construction activity in China; price driven substitution (DCAs)

Source: CRU, Concord Resources Limited, February 2019

CONCORD

11

Statistics typically show zinc metal market running another deficit, albeit reduced, in 2018

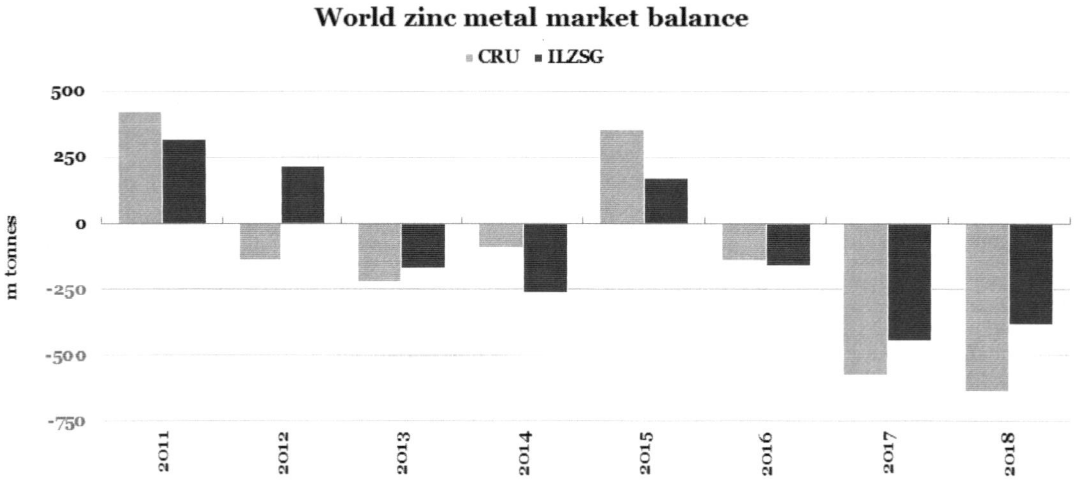

- Struggling zinc supply part offset by soft demand last year, itself owing in part to high prices
- This limited metal market deficit but cumulative shortfall upwards of ~1mt in last three years

Source: CRU, ILZSG, Concord Resources Limited, February 2019

CONCORD 12

Falling exchange stocks frequently flagged as evidence of deficit in the zinc metal market

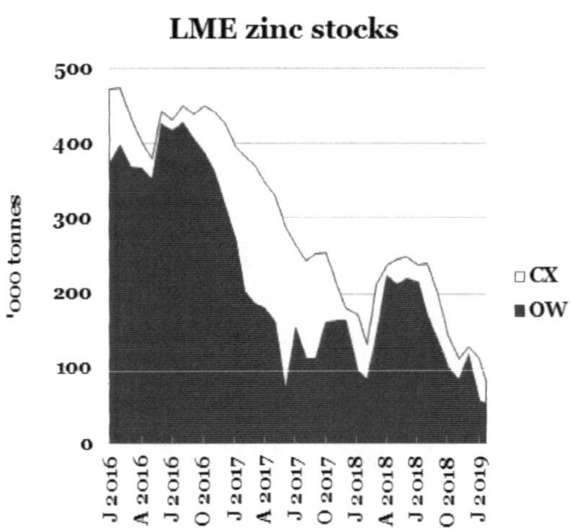

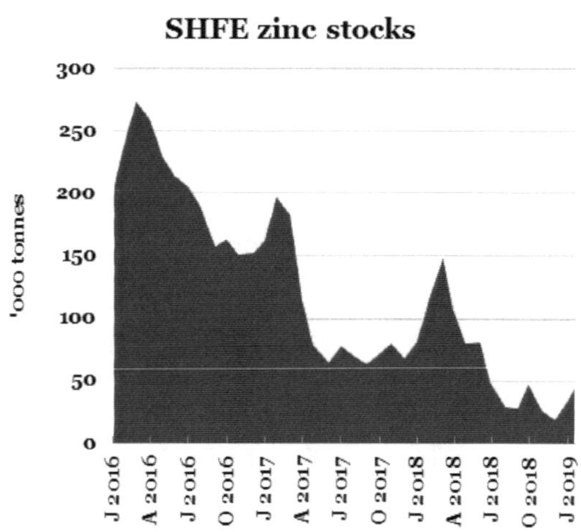

- LME stocks -390kt since Jan 2016, lowest since late 2007; SHFE stocks -160kt, near record low
- Market deficits also buffered by off-exchange stocks from substantial surpluses of early 2010s

Note: month end data or latest
Source: LME, SHFE, Concord Resources Limited, February 2019

CONCORD 13

There has been tightness in the exchange market for zinc; nearby LME spreads have traded at steep backwardation in recent months

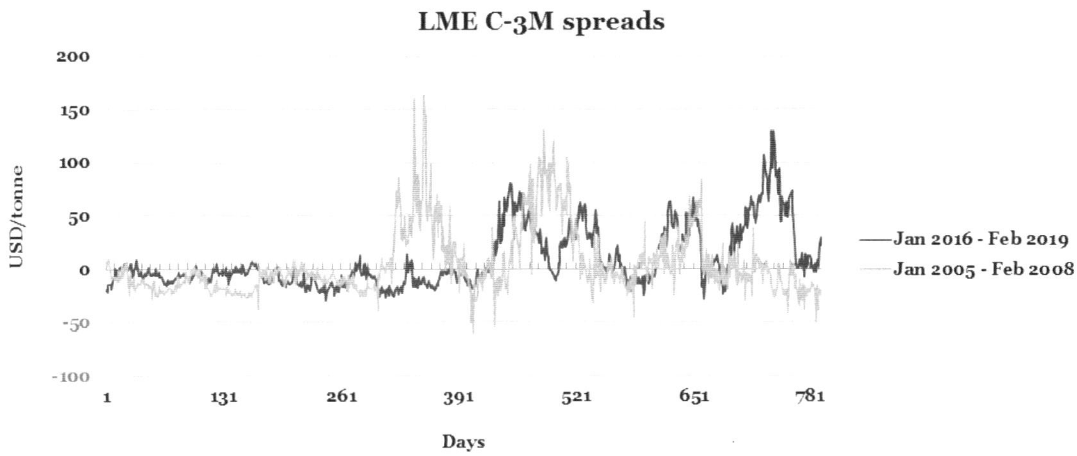

- LME C-3M averaged > USD75/tonne backwardation in Dec 2018, steepest since Dec 2006
- Averages backwardated nine months running since Jun 2018, longer stretch than mid-2000s

Note: all data basis official settlement prices
Source: LME, Concord Resources Limited, February 2019

CONCORD 14

However, there's been less sign of such tightness in the wider market for physical zinc metal

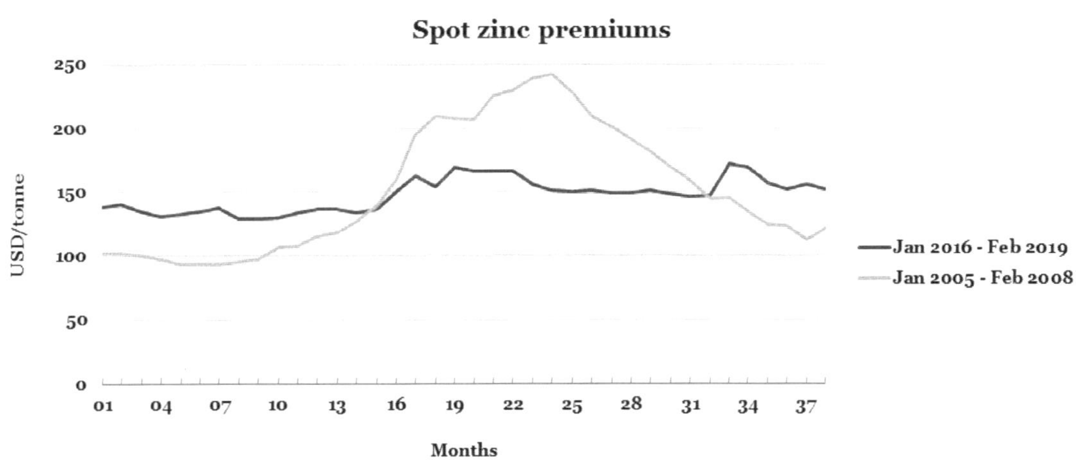

- Average SHG spot premiums up only one-third in recent bull market; Europe especially soft
- In contrast, average spot premiums more than doubled to ~USD250/tonne in late 2006

Note: chart shows average of spot premiums for SHG ingot CIF China; DDP US Midwest; and EXW NW Europe
Source: CRU, Fastmarkets, Concord Resources Limited, February 2019

CONCORD 15

2018-19 zinc market dynamics in three words – strong, struggling, soft

- 2018 market review
- 2019 market outlook
- Summary & conclusions

CONCORD

16

Ex-China zinc smelter output up yet concs supply still sufficient for China to increase imports

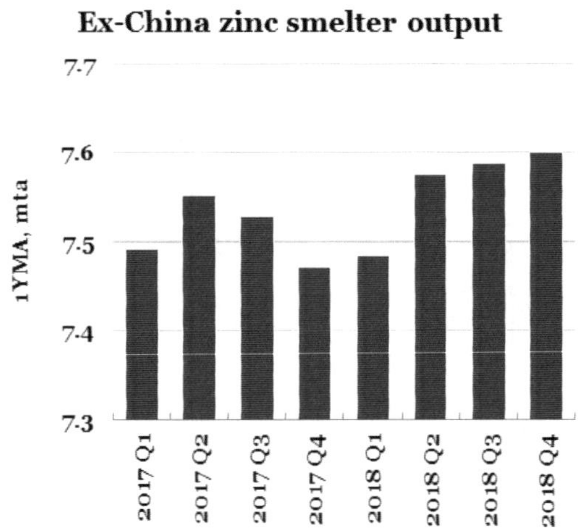

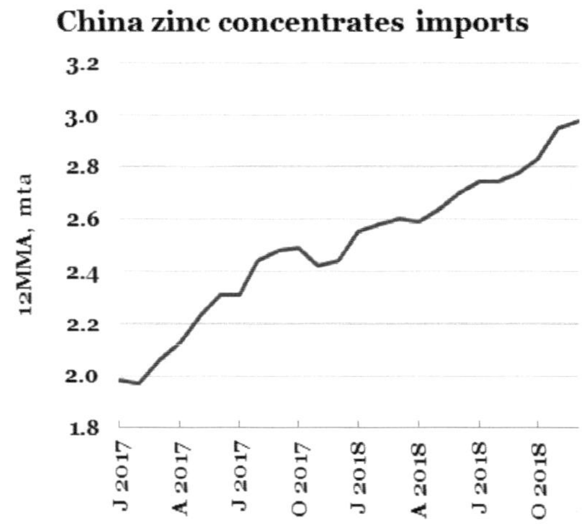

- Ex-China smelter output up +2% YoY to ~7.6mta in Q4 2018 (large part from public listed firms)
- 12MMA China concs imports nearing 3mta; Q4 alone up > 35% YoY, led by AUS, new entrants

Note: China's concentrates import data gross weight basis
Source: China Customs, CRU, Concord Resources Limited, February 2019

CONCORD

17

Higher TCs for zinc concs signal rising supply, easing market balance

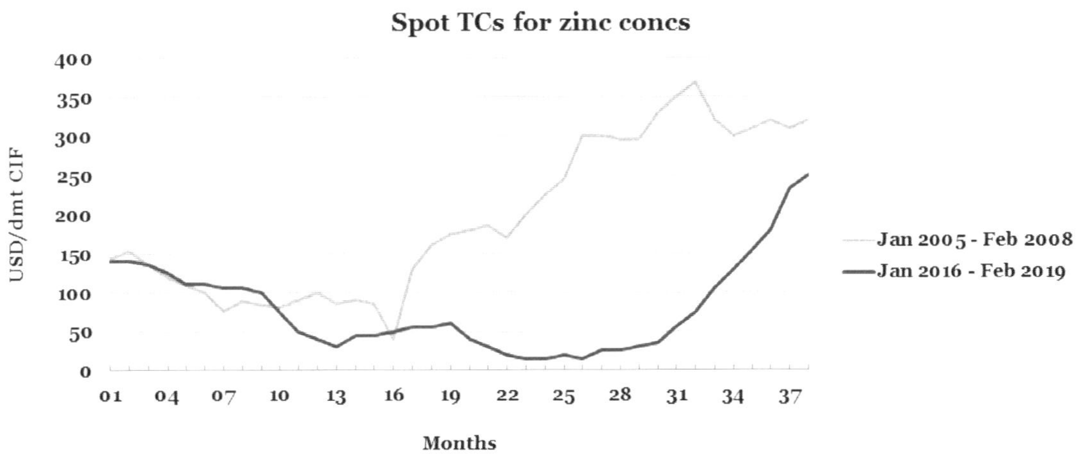

- Spot TCs have surged, albeit from very low levels, in early 2018; latest ~USD250/dmt CIF
- Similar to last cycle: 12m to Feb 2019 up ~USD235/dmt; 12m to Apr 2007 up ~USD255/dmt

Note: TCs for typical clean concentrates ±53% zinc on standard payment terms
Source: CRU, FastMarkets, Concord Resources Limited, February 2019

CONCORD 18

Spot TCs now trading above "benchmark" for the first time since 2007

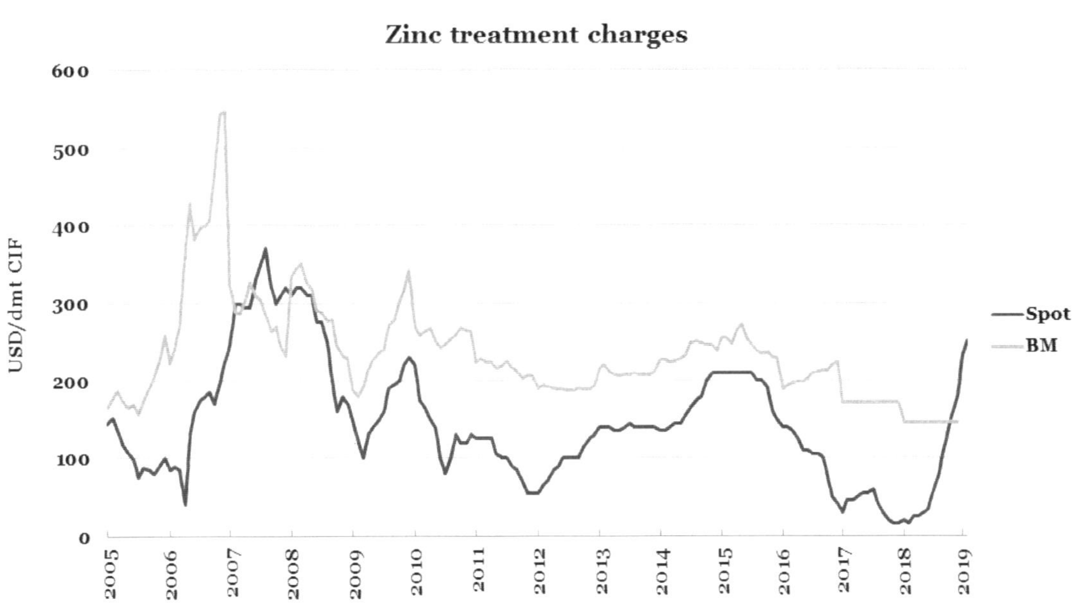

Note: TCs for typical clean concentrates ±53% zinc on standard payment terms. 2019 BM not settled at time of writing
Source: CRU, FastMarkets, LME, Concord Resources Limited, February 2019

CONCORD 19

Zinc smelters' revenues now reaching highest levels in over ten years, miners' revenues still above average of last decade

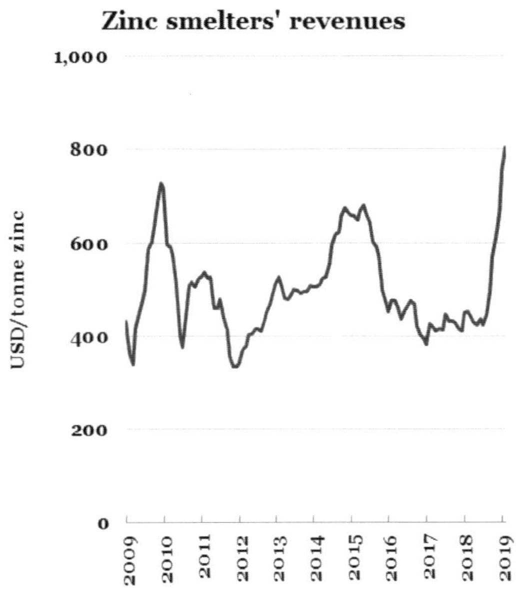

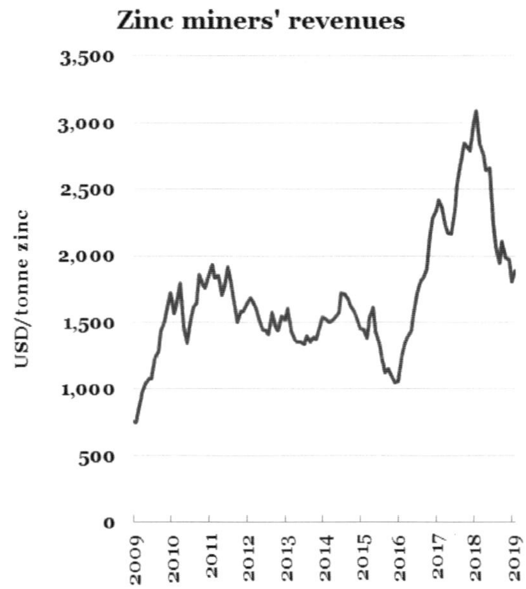

Note: Indicative calculations basis spot TCs for typical clean concentrates ±53% zinc on standard payment terms. By-products, metal premiums not included
Source: CRU, FastMarkets, LME, Concord Resources Limited, February 2019

CONCORD 20

World zinc mine output is forecast to continue rising strongly in 2019 … but there are risks

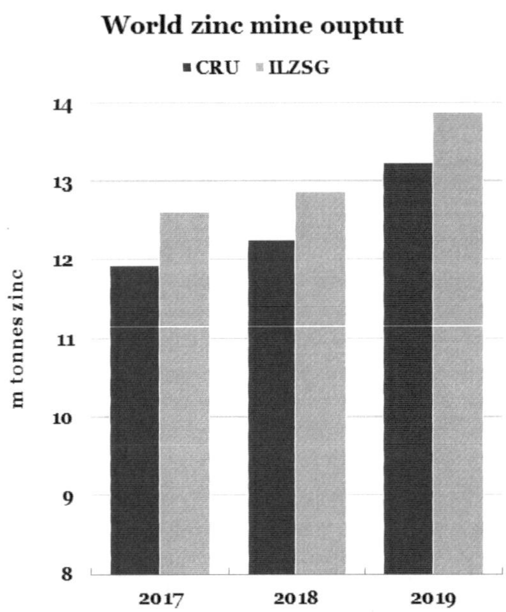

- Leading published forecasts show world zinc mine output rising by approaching 1mt or more in 2019 (+8% YoY)
- Ex-China key contributions expected from (1) Gamsberg, (2) MIM, (3) New Century, (4) Rampura, (5) Penasquito; collectively ±500,000 tonnes
- Nonetheless, meeting forecasts will still need significant increase in output from China
- Uncertainties remain wrt environment, safety constraints in China, as well as project performance ex-China

Source: CRU, ILZSG, Concord Resources Limited, February 2019

CONCORD 21

Flowing through to higher stronger zinc metal output from smelters … but again there are risks

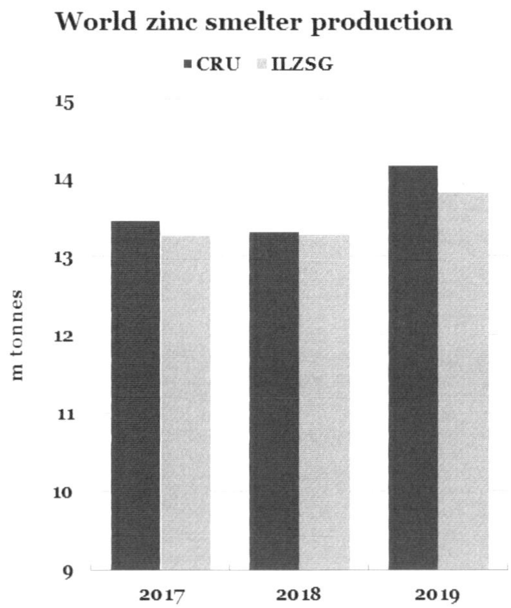

Source: CRU, ILZSG, Concord Resources Limited, February 2019

- Leading published forecasts show world zinc metal output rising by upwards of 500,000 tonnes in 2019
- Forecasts call for key contribution from Chinese zinc smelters
- Financial incentive for zinc smelters to step up production appears strong; latest spot-based revenues > USD800/tonne zinc, highest since H1 2008
- Uncertainties remain wrt environmental, credit constraints on smelters in China
- Chinese EPCM firms report strong orders

CONCORD

22

World zinc consumption is closely correlated with industrial production

- YoY change in industrial production "explains" majority of YoY change in zinc consumption
- Note negative intercept of trendline on y-axis – i.e. zinc consumption typically lags IP

Note: data from 1999-2018 (exc. 2009-2010 (GFC dislocation))
Source: CPB, CRU, Concord Resources Limited, February 2019

CONCORD

23

Good news – world's largest economy is still expanding...

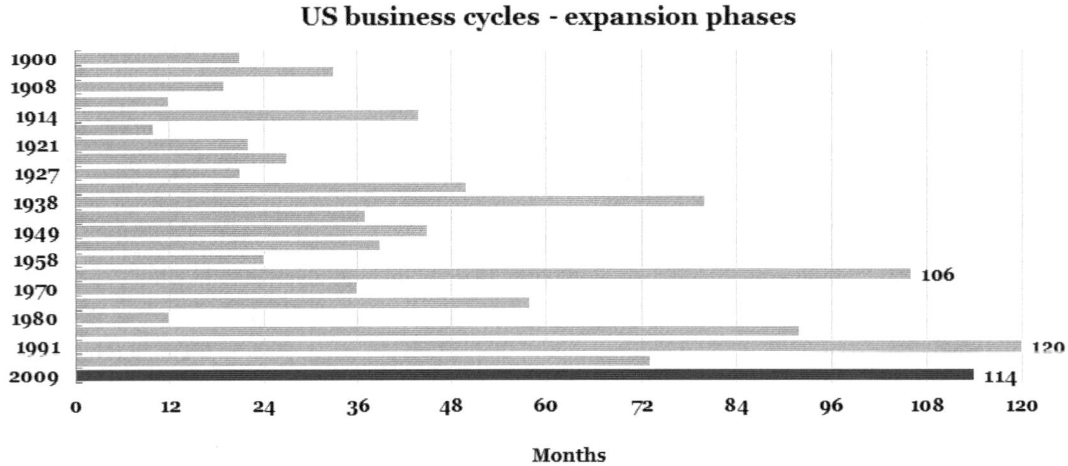

- Yet we have to ask where this leaves balance of risk in the outlook for the business cycle
- And day-to-day experience of the physical zinc metal market is still not reported to be tight

Source: NBER, Concord Resources Limited, February 2019

Latest leading economic indicators point to a significant slowdown in industrial production in the coming months

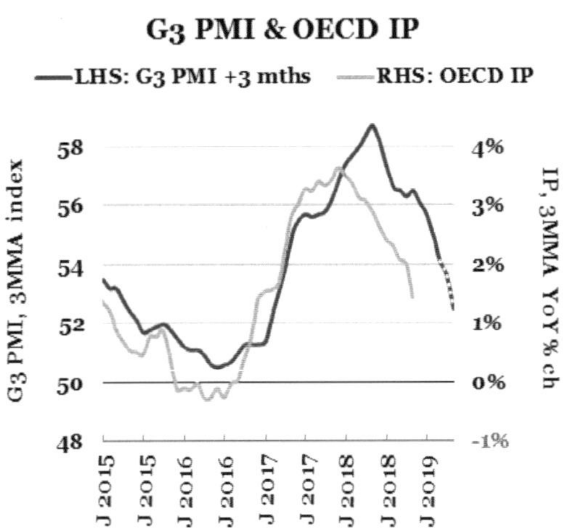

- G3 Feb PMI flash fell -1.3 pts to 51.7 (LR ave 52.2). EZ & Japan < 50 (i.e. output contracting)
- China's Jan NBS & Caixin PMIs both < 50, latter near 3-year low

Note: PMIs for manufacturing. G3 – USA, Eurozone & Japan. G3 PMI calculation weighted according to industrial production
Source: ISM, Markit, NBS, OECD, Concord Resources Limited, February 2019

And world has tipped into a downgrade cycle in economic forecasts; negative for sentiment

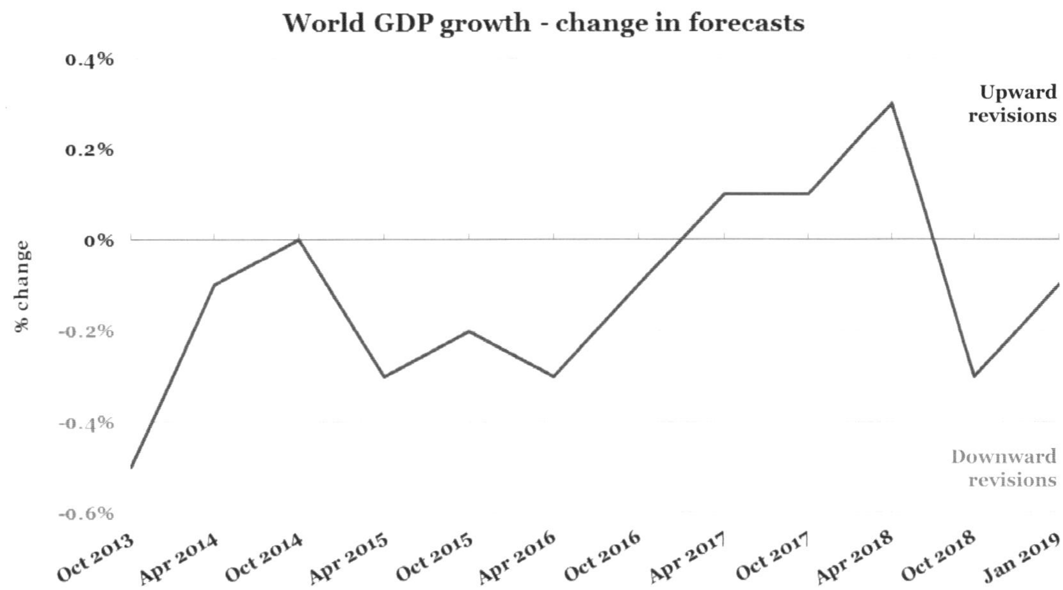

Note: chart shows change in forecast for year following date of issue vs that published in the immediately preceding period; for example, forecast for 2019 in Oct 2018 vs Apr 2018
Source: IMF, Concord Resources Limited, February 2019

CONCORD

World zinc consumption is forecast struggling for growth in 2019

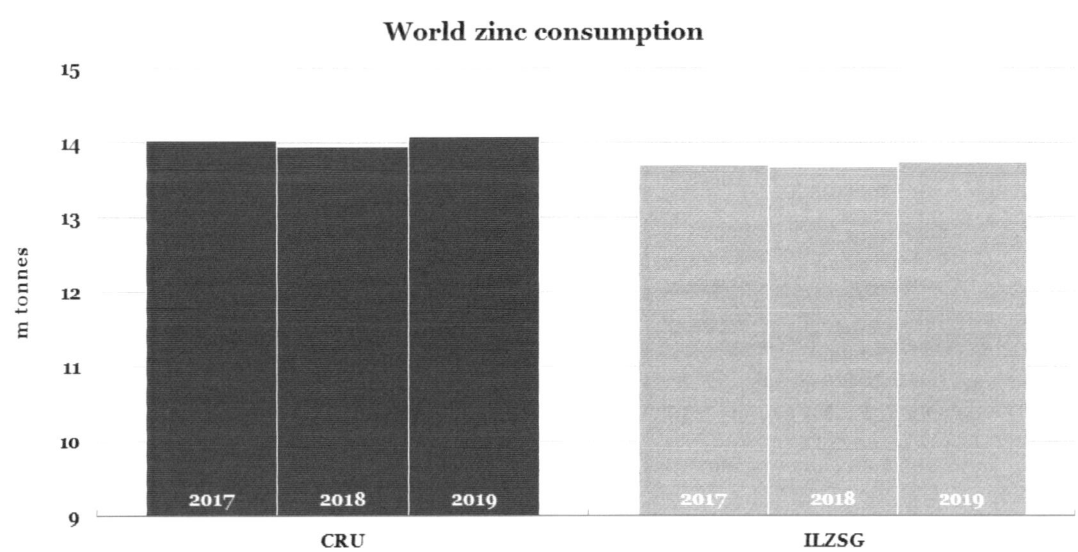

- Sensitivity ±1% IP >> ±140kt zinc consumption; balance of risks skewed to the downside

Source: CRU, ILZSG, Concord Resources Limited, February 2019

CONCORD

2018-19 zinc market dynamics in three words – strong, struggling, soft

- 2018 market review
- 2019 market outlook
- Summary & conclusions

CONCORD

2018-19 zinc market dynamics in three words – strong, struggling, soft

Summary and conclusions

2018 market review

- **Strong** zinc prices; annual average at 11-year high
- **Struggling** zinc mine output still below 2014-15 peak, metal production down for 3rd year running; both owing largely to China
- **Soft** zinc consumption, down last year for first time since GFC, again owing largely to China

2019 market outlook

- **Strong** forecasts for zinc mine output and metal production, subject to uncertainties in China
- **Struggling** zinc consumption, with leading indicators points to slowdown in world industrial production
- **Softer** zinc prices under pressure from combination of rising supply, slowing consumption

Source: Concord Resources Limited, February 2019

CONCORD

2018-19 zinc market dynamics in three words – strong, struggling, soft

- **Thank you for your attention!**

- **Questions welcome to:**

  Duncan Hobbs

  +44 20 3879 3966

  duncan.hobbs@concordltd.com

CONCORD

Concord Resources Limited

Our markets: every step of the supply chain

	Aluminum	Copper	Zinc, lead, nickel	Alumina, bauxite	Precious metals
Metals				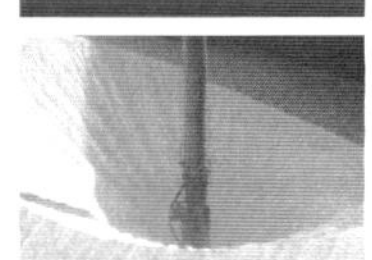	
Products	• P1020 • Billets • Alloys • Rod	• Cathode • Copper rod • Concentrates	• SHG & CGG zinc • 397 & 49s lead • Primary nickel • Concentrates	• SGA • Raw bauxite	• Gold • Silver • PGMs
Origination	• Consignments • Prepays/financing • Off-take/marketing • Tolling	• Consignments • Prepays/financing • Off-take/marketing • Tolling	• Consignments • Prepays/financing • Off-take/marketing • Tolling	• Swaps • Prepays/financing • Off-take/marketing • Tolling	• Off-take/marketing • By-products from concentrates
Storage / logistics	• LME system • Warehousing • Shipping	• LME system • Warehousing • Shipping • Blending	• LME system • Warehousing • Shipping • Blending	• Warehousing • Shipping • Blending	• Vaulting
Consumers	• Aerospace • Automotive • Construction • Packaging	• Wire and cable • Tubing • Brass • Smelters	• Galvanizers • Battery makers • Stainless & alloy steelmakers	• Refineries • Smelters	• Banks • Smelters

CONCORD

Concord Resources Limited

Contact our marketing teams

New York	London	Hong Kong
437 MADISON AVENUE 28TH FLOOR NEW YORK, NY 10022 USA	3 ST JAMES'S SQUARE LEVEL 8 LONDON, SW1Y 4JU ENGLAND	41 CONNAUGHT ROAD CENTRAL LEVEL 16 NEXXUS BUILDING HONG KONG
Scott Evans Senior Trader Americas Tel: +1 212 822 9793 scott.evans@concordltd.com	David Freeland Senior Trader EMEA Tel: +44 20 3879 3995 david.freeland@concordltd.com Kareem Barbir Senior Trader Concentrates Tel: +44 20 3879 3993 kareem.barbir@concordltd.com	Felix Cauro Senior Trader Asia Tel: +852 3757 9915 felix.cauro@concordltd.com

CONCORD

32

IHS Markit

Zinc Oxide Market Overview

International Zinc and Zinc Oxide Conferences | Scottsdale, Arizona

February 2019

Heike Held, PhD

heike.held@ihsmarkit.com

IHS

Zinc Oxide:

Global Capacity and Production

Consumption

- By region
- By end use
  - Rubber compounding
  - Ceramics and glasses
  - Agriculture
  - Chemicals
  - Paints and coatings
  - Varistors

IHS Markit 3

Global Zinc Oxide Capacity and Production

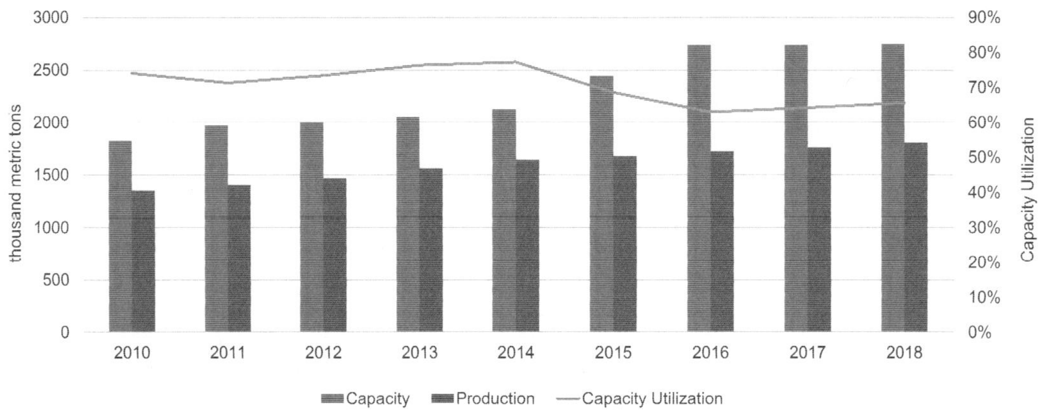

IHS Markit 4

2018 Capacity Utilization by Region

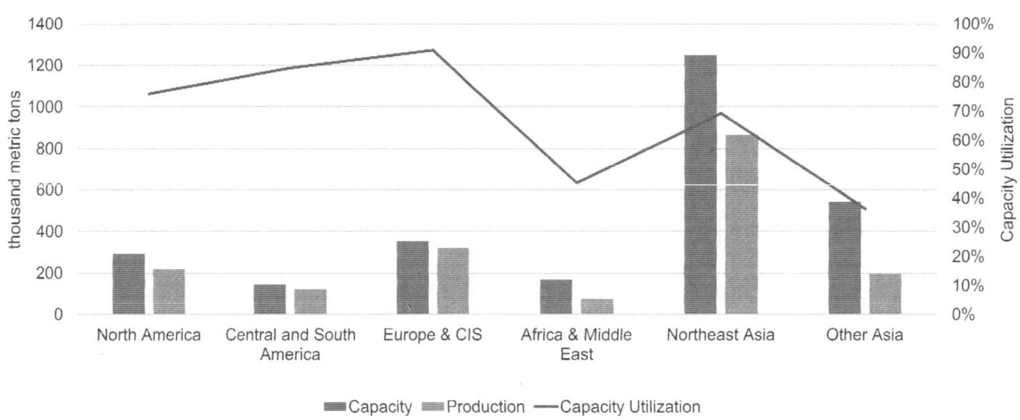

2018 Zinc Oxide Consumption by Region

Total global consumption: 1.8 million metric tons

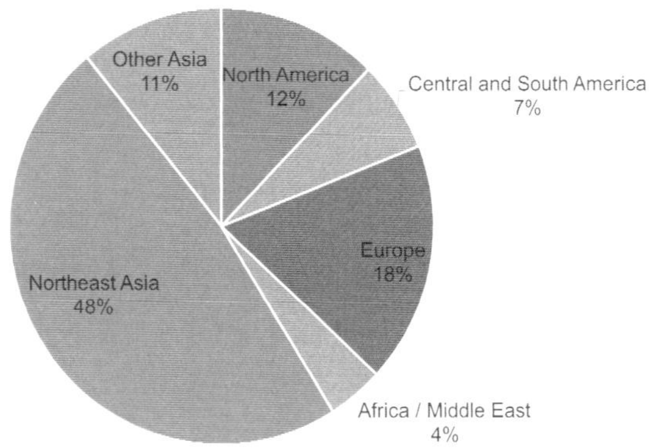

2018 Main Uses of Zinc Oxide

Total global consumption: 1.8 million metric tons

- Rubber compounding remains major end use in all regions
- Ceramics & Glasses is second largest application in all regions but North America

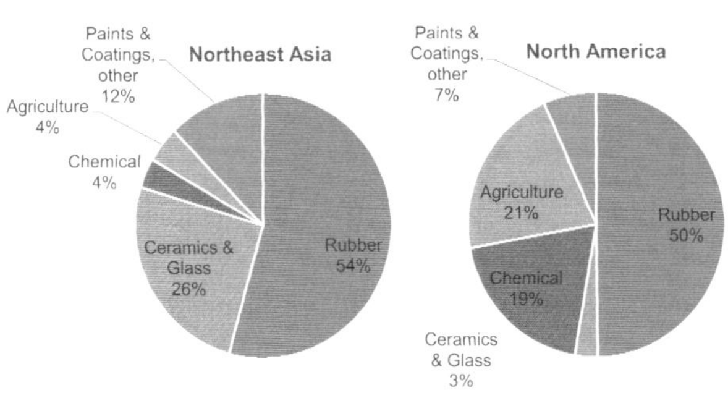

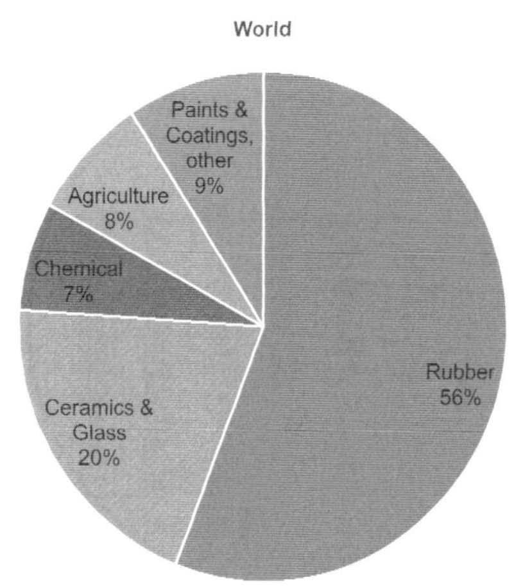

IHS Markit 7

2018 ZnO Consumption in Rubber Compounding by Region

Total consumption in the rubber segment: 1.0 million metric tons

- Approximately 75-80% of rubber is used in tire production,
- i.e. 42-45% ZnO used in tire production.

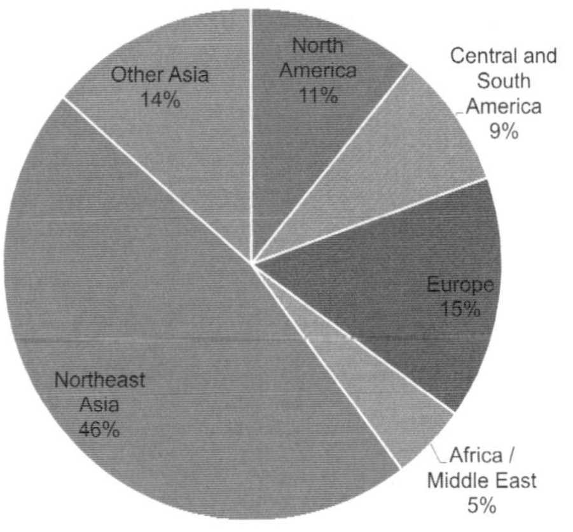

IHS Markit 8

Consumption of Polybutadiene Rubber (PBR) in Tire Production

Largest percentual growth expected in India, Southeast Asia

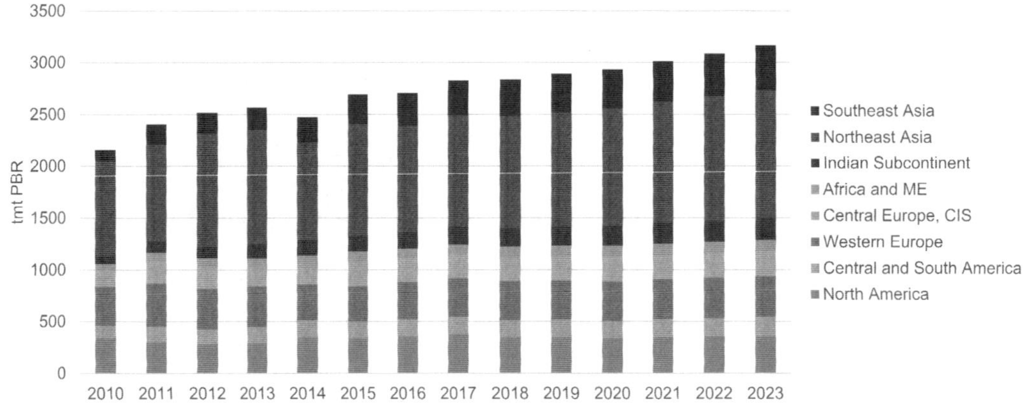

Does ZnO consumption correlate with vehicle production?

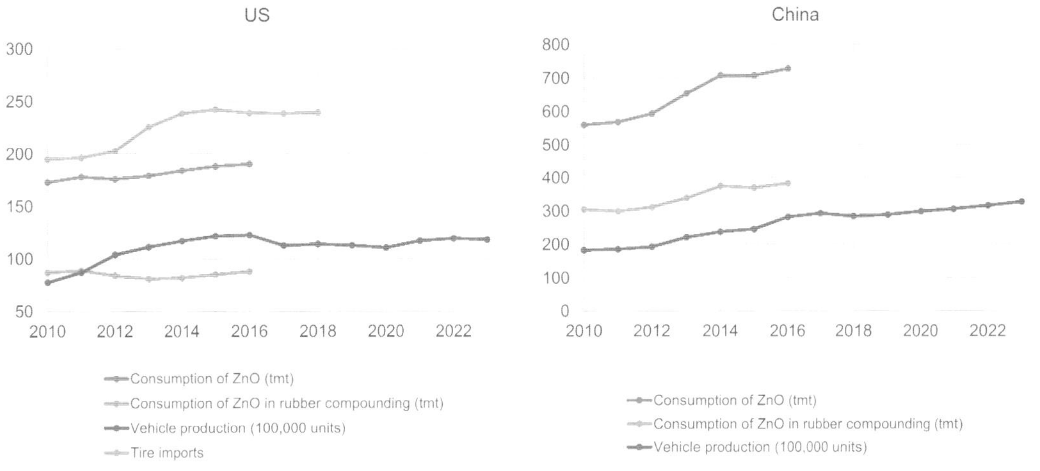

Consumption of PBR in the Production of General Rubber Goods

Largest percentual growth expected in CIS, Africa

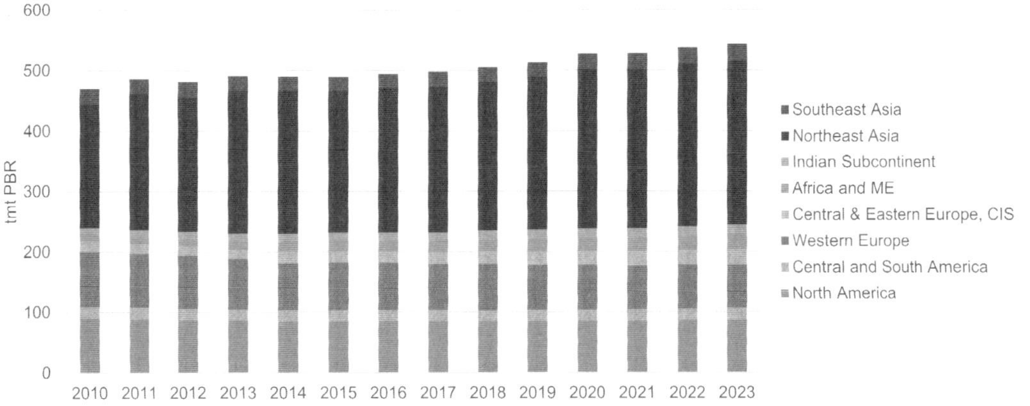

2018 ZnO Consumption in Ceramic Tiles Production by Region

Total global consumption in the ceramics segment:

360 thousand metric tons

- China is still top ceramic tile producer, in spite of recent reduction of production
- Reduction in tile production in Indonesia is offset by production increase in India and Vietnam

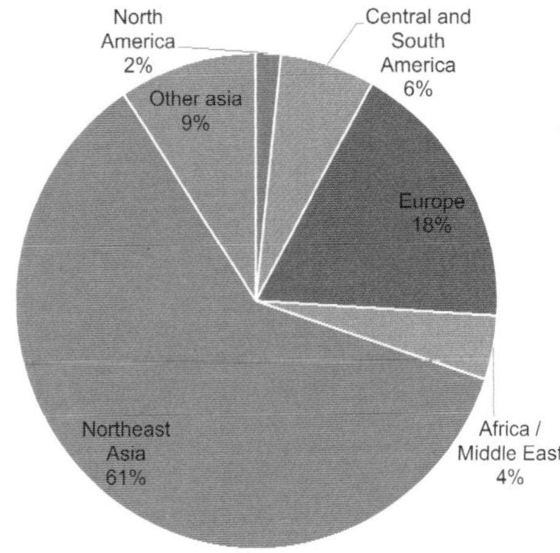

Production of Ceramic Tiles

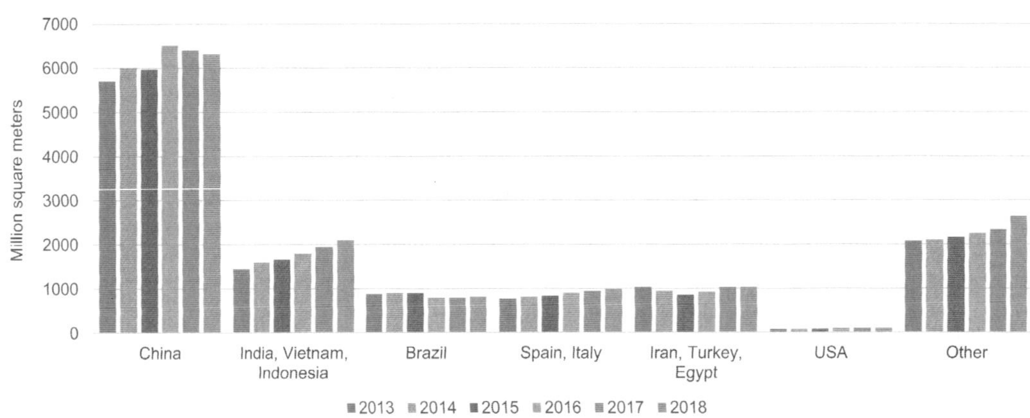

Extra-regional Trade of Frits containing 12 – 20% ZnO

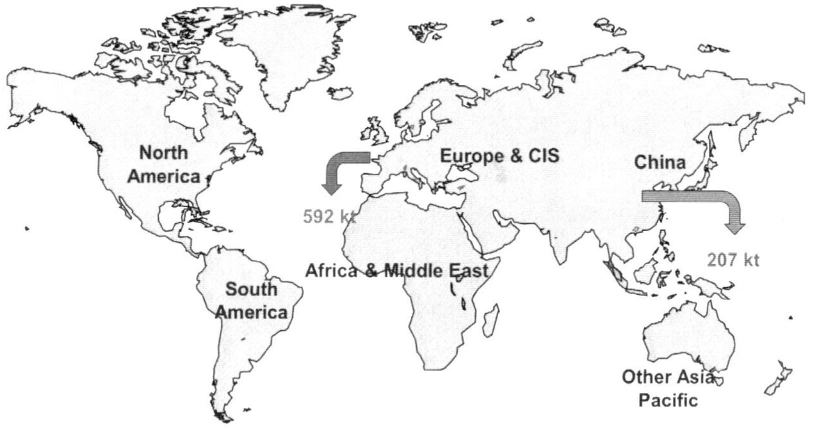

2018 Consumption of ZnO in Agriculture

Total consumption in agricultural uses: 142 tmt ZnO (114 tmt Zn)

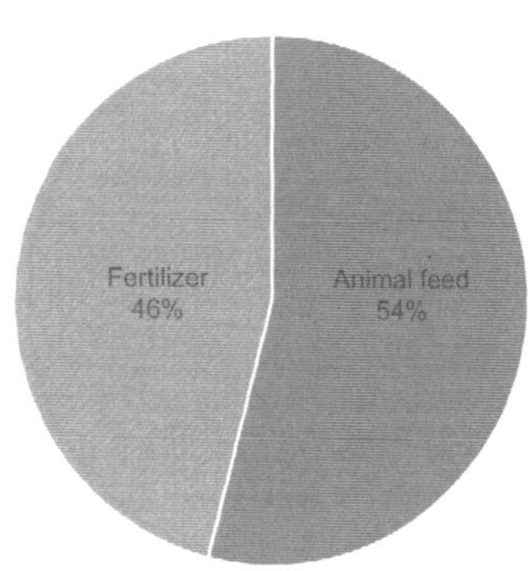

World Consumption of Zinc Chemicals in Agriculture

Zn is used mostly in the form of Zinc Sulfate.

Total consumption of Zn in agriculture: 390 tmt

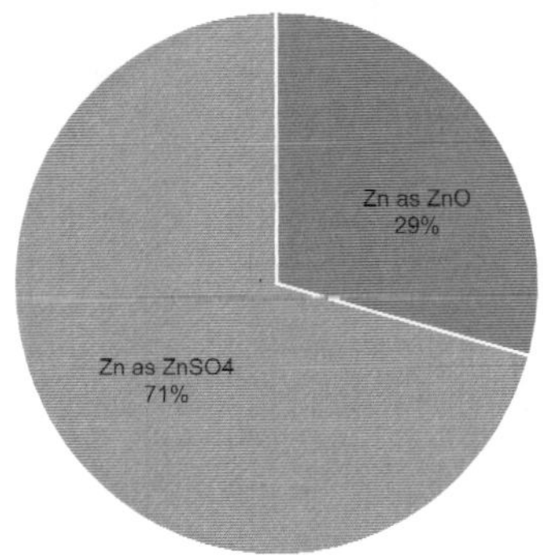

2018 Consumption of ZnO in Chemical Uses by Region

Total consumption in chemical uses: 124 tmt

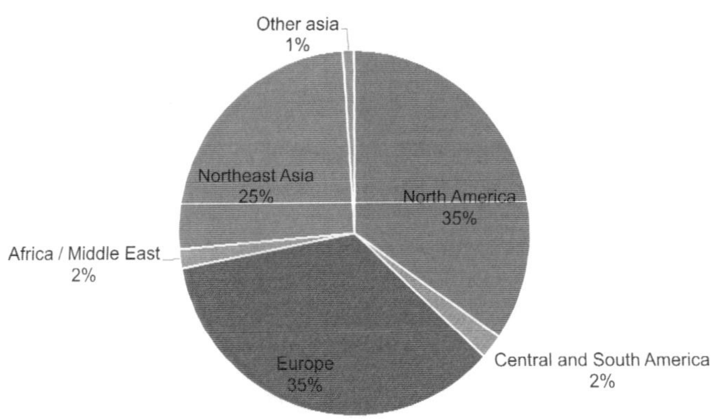

IHS Markit 17

ZnO in Chemical Uses

Growth in important areas within chemical uses:

- Catalysts in desulfurization (oil drilling and refining, chemical manufacturing) → environmental protection
- preparation of
    - Zinc Phosphates (ZDDP) (lubricant additives) → energy conservation
    - Zinc stearate
    - Zinc octoate
    - Zinc bromide (drilling fluids)

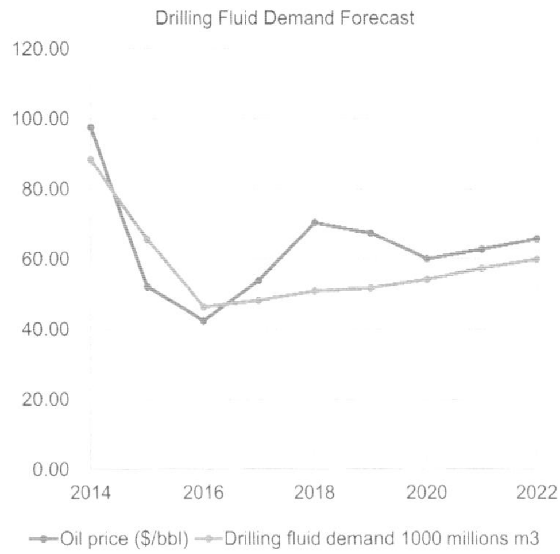

IHS Markit 18

ZnO in Paints and Coatings

Slightly growing markets in specialty areas of use, i.e.

- Antifouling
- Antimicrobial
- UV-protection
- Mechanical Strength of the Coating

Zinc Oxide in Varistors

- Consumption of ZnO in varistor production in 2015 close to 50 ktpa.
- Most of this being consumed in China.
- Strong growth expected to continue.
- Growth Drivers:
  - Continuing growth of handsets.
  - Automotive electronics growth
  - Growth of arresters with growth of smart grid technology
  - Renewable energy systems grow and require redundant circuit protection

Conclusion:

Demand for ZnO is expected to continue to rise over the next five years.

Key demand drivers for ZnO:

- Growing automotive market in Asia
- Global increase in tile manufacturing, in spite of receding production in China
- Growing importance of zinc as micronutrient; agricultural programs especially in Asia, Africa and South America.

THANK YOU!

Time to concentrate on zinc smelters

Wood Mackenzie
A Verisk Business

Jonathan Leng

Trusted Intelligence

woodmac.com

A tumultuous year for zinc

Trade wars and market uncertainty drove zinc market sentiment in 2018

Soaring to over $3600/t before plunging to less than $2300/t

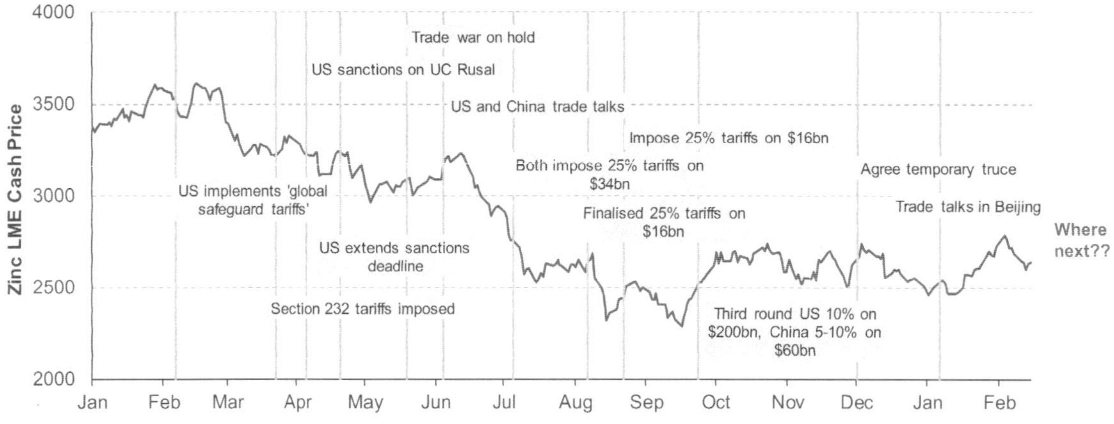

Source: Wood Mackenzie

woodmac.com

Outlook for regional zinc consumption growth

Slower demand growth still equates to 230kt/a Zn

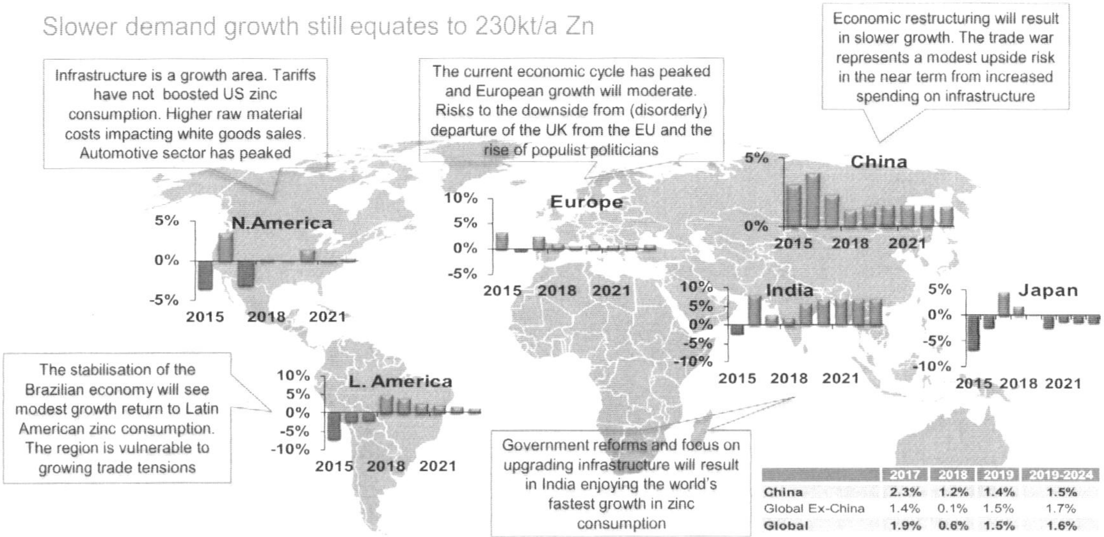

	2017	2018	2019	2019-2024
China	**2.3%**	**1.2%**	**1.4%**	**1.5%**
Global Ex-China	1.4%	0.1%	1.5%	1.7%
Global	**1.9%**	**0.6%**	**1.5%**	**1.6%**

Source: Wood Mackenzie

3

woodmac.com

Forecast mine production capability to rise by 2.5Mt/a Zn by 2021

Plus potential additional production from mine life extensions and projects

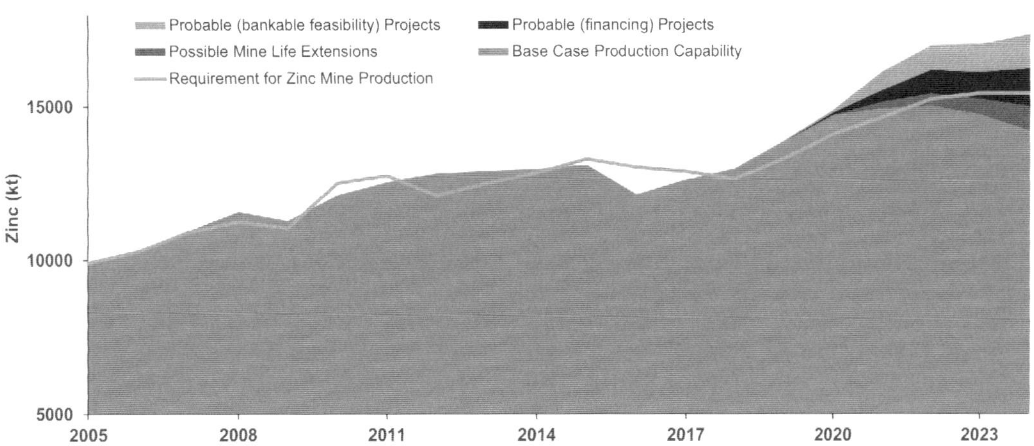

Source: Wood Mackenzie

4

woodmac.com

No shortage of advanced stage mine projects

Zinc price expectations and availability of financing are critical to incentivise development of Probable status projects

Historical mine starts-ups and future projects

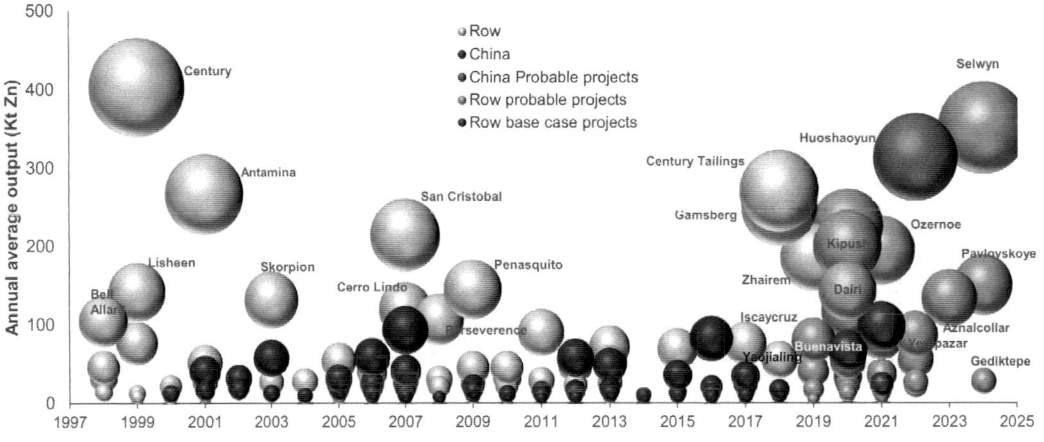

Source: Wood Mackenzie

woodmac.com

Miners not under margin pressure

At both M1 and M2 margin

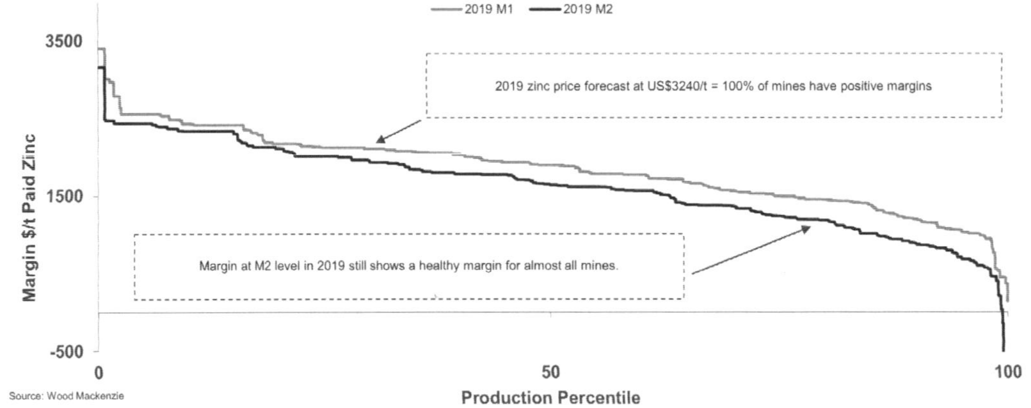

Source: Wood Mackenzie

Deleterious elements in new mine supply – a manageable issue for smelters

Blend to smelter limits or install smelter process technology

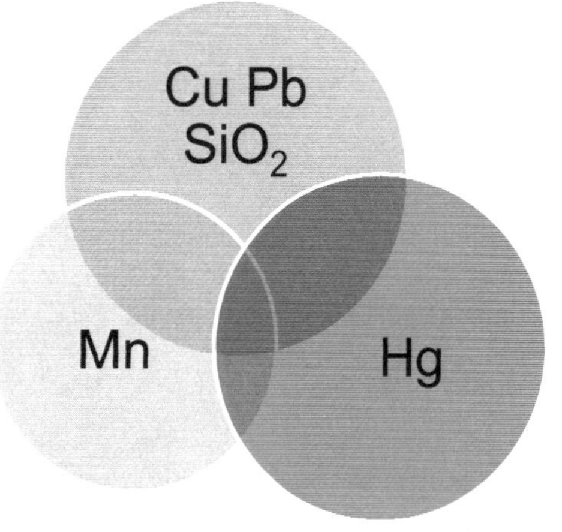

Source: Wood Mackenzie

woodmac.com

Chinese and rest of world concentrate markets have moved to surplus

Strong growth in mine supply in the next couple of years will lead to a large concentrate surplus

Concentrate market balance

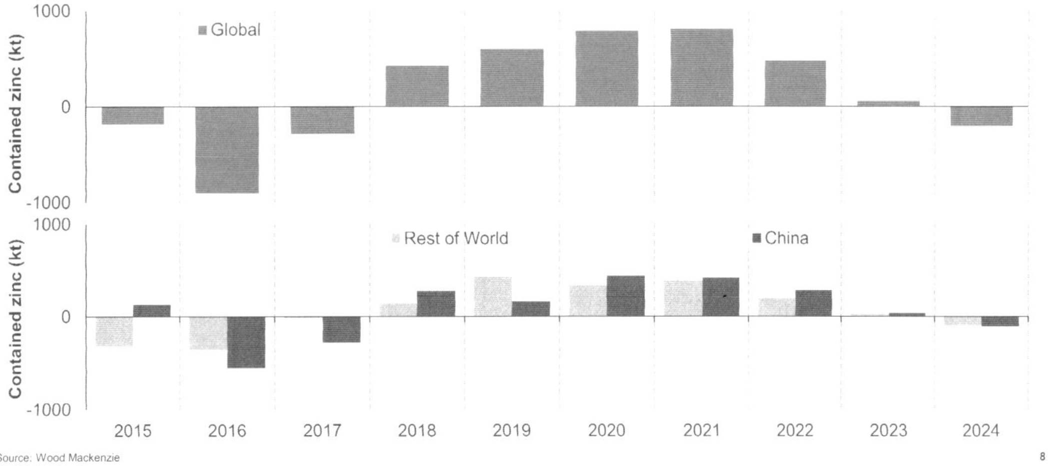

Source: Wood Mackenzie

8

Zinc spot TC's have climbed rapidly

China spot TCs up from $24/t in H1 2018 to over $200/t flat CIF MCP in early-2019
Benchmark TC highly likely to be settled higher for 2019

Source: Wood Mackenzie

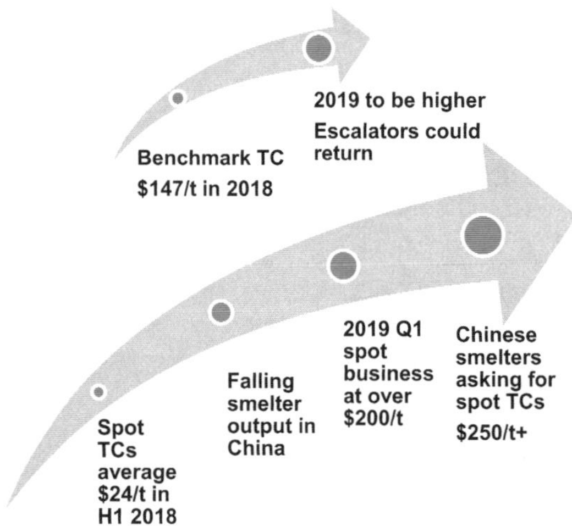

woodmac.com

Smelter revenues from zinc concentrate to increase in 2019

Higher TCs to bolster smelter margins, plus greater bonus zinc

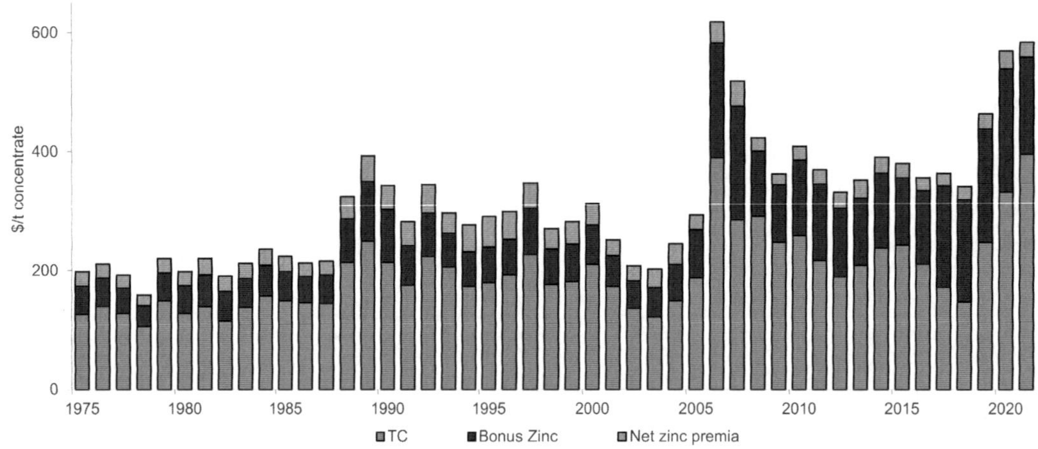

Source: Wood Mackenzie

10

woodmac.com

The increased availability of concentrate should enable an increase in utilisation rates

But with utilisation rates already high in the rest of the world the driver will have to be China

Smelter utilisation rates

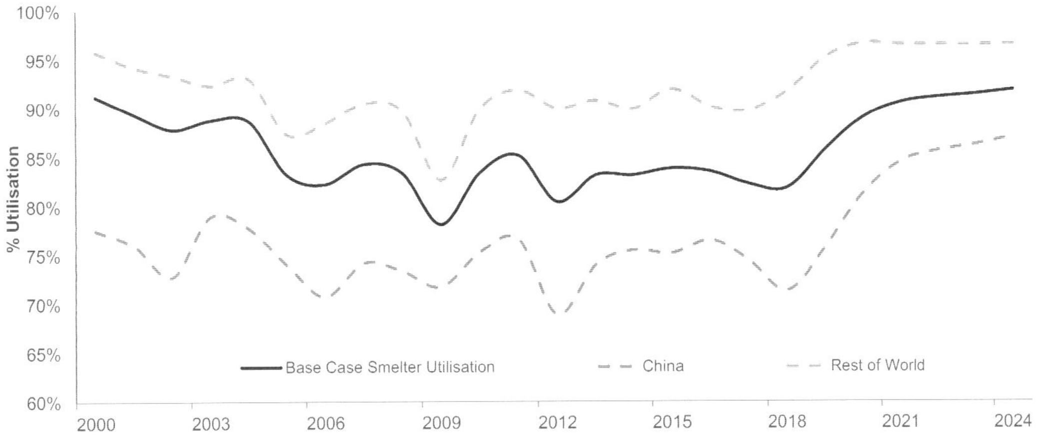

Source: Wood Mackenzie

11

woodmac.com

The crackdown on pollution has spread beyond the 2+26

The enforcement of environmental regulations has spread beyond mining into the smelting sector – resulting in production cuts and delays in the commissioning of new capacity

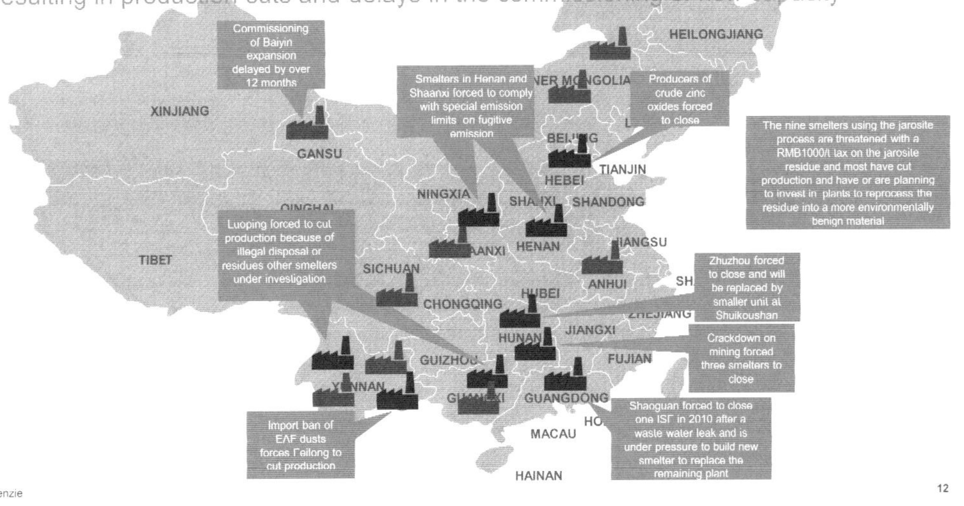

Source: Wood Mackenzie

12

woodmac.com

Exchange stocks fall to less than four days of global consumption

Zinc price volatility increases once stocks fall below the equivalent of 10 days of global consumption. This was breached in June 2017 and continued to fall to less than 4 days of consumption in by the end of 2018

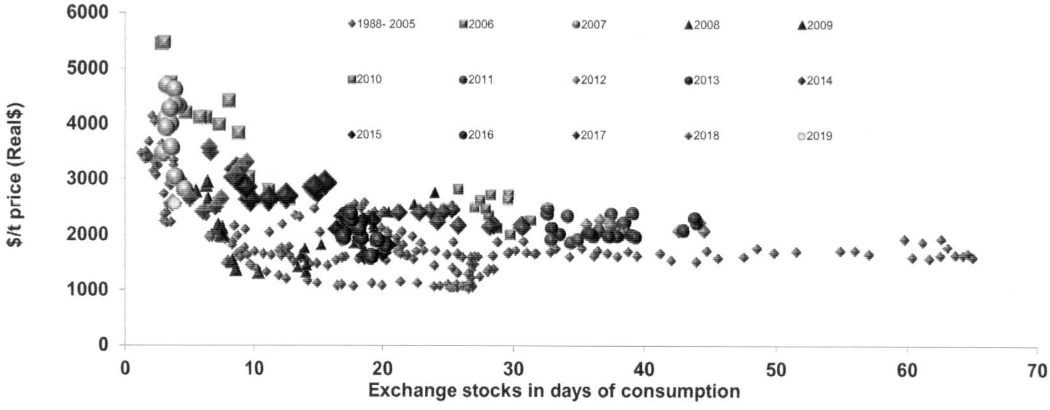

Source: Wood Mackenzie

13

woodmac.com

Falling metal stocks provides fundamental support to higher zinc prices

Peak price forecast of US$3700 reflects refined market tightness before the downturn in 2021

Refined zinc balance and price, 2000-2024

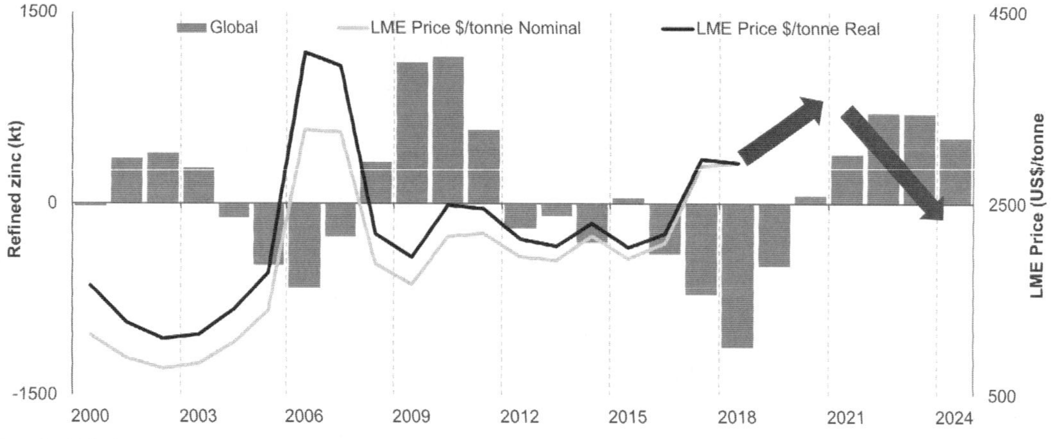

Source: Wood Mackenzie

14

Key takeaways

- Strong growth in mine supply in the next couple of years is expected to lead to a large concentrate surplus
- The ability of Chinese smelters to increase capacity utilisation is a critical risk
- With refined metal demand expected to grow, there is a requirement for added zinc smelter capacity
- Low refined stocks will provide fundamental support to prices

Jonathan Leng

Principal Analyst Zinc Markets

Biography

Jonathan is a principal analyst focused on zinc mine and smelter supply, plus concentrate analysis. Prior to his current role, Jonathan was manager of the zinc, lead and gold mine cost research teams. He has 20 years of metals and mining experience, primarily in base metal, ferroalloy and gold research.

Prior to joining Wood Mackenzie, he worked as a geologist for Croesus Gold in Australia and metals research at CRU International. Jonathan also worked in construction cost management at Davis Langdon.

Jonathan holds a M.Sc. Mining and Mineral Resource Evaluation, B.Sc. (hons) Geology and P.G.Dip. in Construction Cost Management.

Connect with Jonathan

Jonathan.Leng@woodmac.com

+44 1483 596 626

woodmac.com

Disclaimer

Strictly Private & Confidential

- These materials, including any updates to them, are published by and remain subject to the copyright of the Wood Mackenzie group ("Wood Mackenzie"), and are made available to clients of Wood Mackenzie under terms agreed between Wood Mackenzie and those clients. The use of these materials is governed by the terms and conditions of the agreement under which they were provided. The content and conclusions contained are confidential and may not be disclosed to any other person without Wood Mackenzie's prior written permission. Wood Mackenzie makes no warranty or representation about the accuracy or completeness of the information and data contained in these materials, which are provided 'as is'. The opinions expressed in these materials are those of Wood Mackenzie, and nothing contained in them constitutes an offer to buy or to sell securities, or investment advice. Wood Mackenzie's products do not provide a comprehensive analysis of the financial position or prospects of any company or entity and nothing in any such product should be taken as comment regarding the value of the securities of any entity. If, notwithstanding the foregoing, you or any other person relies upon these materials in any way, Wood Mackenzie does not accept, and hereby disclaims to the extent permitted by law, all liability for any loss and damage suffered arising in connection with such reliance.

17

woodmac.com

Verisk™

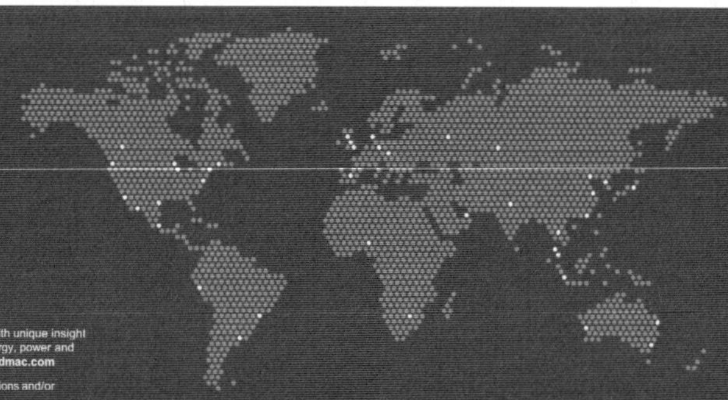

Europe	+44 131 243 4400
Americas	+1 713 470 1600
Asia Pacific	+65 6518 0800
Email	contactus@woodmac.com
Website	www.woodmac.com

Wood Mackenzie™, a Verisk business, is a trusted intelligence provider, empowering decision-makers with unique insight on the world's natural resources. We are a leading research and consultancy business for the global energy, power and renewables, subsurface, chemicals, and metals and mining industries. **For more information visit: woodmac.com**

18

Zochem

ZnO

Zinc Oxide Market Opportunities & Challenges:

Connecting demand with Cultural and Geographical Shifts

Mohit Sharma

February 27, 2019

Scottsdale, Arizona, U.S.A.

Zochem

ZnO

Introduction

Zochem:

- Est. 1974; Acquired by ZO LLC in 2017
- Industry consolidation and amalgamation of strengths of two formidable teams in zinc oxide market intelligence and lean manufacturing
- Supplier of high quality, socially responsible zinc oxide
- Facilities in Dickson, TN and Brampton, ON

Speaker: General Manager of Zochem ULC

- 15+ years in Automotive, Industrial/Chemical Manufacturing
  - Previously with Toyota Motor Manufacturing, General Motors
- Primary Education:
  - B.A.Sc Chemical Engineering (University of Waterloo)
  - M.B.A (University of Western – Ivey School of Business)

Zochem

Economic Overview: Asia-Pacific

Asia-Pacific Region remains the leader in annual GDP growth expectations; but a China "slowdown" is counterbalanced by Vietnam, Indonesia, and Thailand.

Source: World Bank

**China GDP by quarter shown below*

India GDP at 7.7% (Oct 2018) to a 7.3% projection. Source: Trading economics

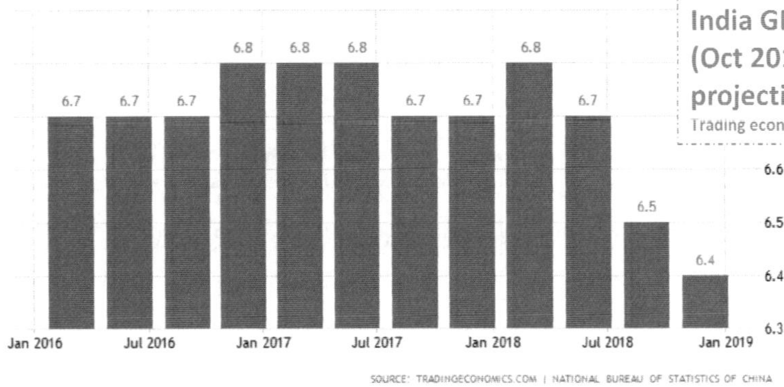

China's full CY2018 ~6.6% GDP lowest since 1990. Source: CNBC, Jan 2019

Zochem

Economic Overview: Europe

Europe and Central Asia showing continued growth in consumption

Sources (OECD, World Bank, IMF, UN) had E.U. 2018 GDP at 1.8%-2.3%; with CY closing at ~1.8%. Same sources had forecasted a range of 1.7%-2.1% for 2019

https://knoema.com/mewdmh/european-union-gdp-growth-forecast-2018-2020-data-and-charts

Growth rates of GDP in volume
(based on seasonally adjusted data)

	Percentage change compared with the previous quarter				Percentage change compared with the same quarter of the previous year			
	2018				2018			
	Q1	Q2	Q3	Q4	Q1	Q2	Q3	Q4
EA19	0.4	0.4	0.2	**0.2**	2.4	2.2	1.6	**1.2**
EU28	0.4	0.5	0.3	**0.3**	2.3	2.2	1.8	**1.5**

Source: Eurostat

Working age population is expected to decrease ~0.5% between 2018 – 2027 (World Bank) which is one of the drivers related to declining GDP projections for the E.U. in the next decade

Source: World Bank

Zochem

D.P.E.S.T.O Analysis - ZnO

- Demographics
  - Millenial population increase (Born 1984-2004) as percentage of total consumer population.
    - Buying habits, purchase decisions and how those impact zinc oxide end use products
      - *Study by University of Akron on tires purchases specifically*
- Political
  - Global trade concerns and barriers are real.
    - U.S. – China ongoing discussions; Tariff threats. Global impact, not just 2 nations
    - Brexit – looming
      - Driving concerns on E.U. business climate index. Germany (strong E.U. example) at lowest score since 2014. 98.5 to a 99.0 forecast.
    - North America (USMCA) ratification – not complete
- Economic
  - U.S. Growth. 2018Q2 GDP ~4.2%; Q3 GDP was ~3.5%. Q4 ~2.7% (11 sources)
    - Has the peak been observed?
  - Freight/Transportation. Tied to above GDP and fuel prices, a tapering off of freight cost increases has been noted in North America specifically.

Zochem

D.P.E.S.T.O Analysis Cont'd

- Social
  - Urbanization has increased from 75% to 82% in North America in the last 30 years.
    - Working from home, less face-to-face meeting, public transit, and environmental awareness
    - https://data.worldbank.org/indicator/SP.URB.TOTL.IN.ZS?locations=XU; UWE/Oxford studies
  - Millenial focus on social responsibility and environmental awareness (tire/oil replacement interval)
    - 10% less teens (age 16-19) getting licenses; driving 1700 miles less annually on average
      - Source: U.S. Energy Information Agency
  - Developing countries increased education on health and wellness (ZNI)
- Technological
  - Electric vehicles, self-autonomous vehicles (tech device vs. just a car) – tie into Demographic
    - No motor oil, chemical additive
  - Motor oil improvements for increased intervals (coincides with oil change indicator)

Zochem

ZnO

Zinc Oxide Industry Size-Up

- Global consumption remains strengthened by 3 predominant industries
  - Tire, Chemical, and Agriculture
  - Other usage: Paint/Coatings, Electronics, Pharmaceutical, Ceramics

- Regional NA Industry has an approximate:
  - Market Size = ~190,000mt
    - 45% Tire
    - 24% Chemical
    - 9% Rubber
    - 8% Agriculture
    - 6% Distribution
    - <3% each (coatings, electronics, plastics, ceramics)
    - **Note: All French Process zinc oxide**
- Recent M&A activity with 2 manufacturers sold within the last 90 days

Zochem

ZnO

Tires

- Overall North American market
  - 70% after market; 30% OEM

US SALES BY TOP AUTOMAKERS

				Year to date		
	October 2018	October 2017	% change	2018	2017	% change
GM	N/A	N/A		2,168,808	2,195,502	▼1.2
Ford	192,616	200,436	▼3.9	2,080,241	2,133,895	▼2.5
Toyota	191,102	188,434	▲1.4	2,015,339	2,019,915	▼0.2
FCA (Chrysler)	177,391	153,373	▲15.7	1,857,374	1,732,511	▲7.2
Honda	122,182	127,353	▼4.1	1,329,179	1,358,956	▼2.2
Nissan	109,962	123,012	▼10.6	1,234,644	1,319,253	▼6.4
Hyundai/Kia	98,127	97,407	▲0.7	1,051,870	1,067,077	▼1.4

NA - not available; GM no longer publicly shares monthly sales data.
Source: Compiled by Fastmarkets from company reports.

OEM auto sales are not the majority of the tire production market, but a signal in the overall economy

- New tire plant additions will drive an increase in demand locally in NA:

Hankook	Giti Tire	Sentry (2020-21)
Nokian (2020)	Yokohama	Kumho (2018)

- Incremental demand vs. cannibalization?
  - Onshore of foreign tire manufacturing is noted to predominantly replace tires currently being imported into the U.S. for both OEM and after market consideration. *RESULT: Increased demand for ZnO in NA*

Zochem

Specialty Chemical

- Motor oil usage (ZDDP)
  - Since 2013, average oil change interval moved between 7,500 and 10,000 miles; more than 2x the historical 3,000 mile norm (Source: Edmunds)
    - Average miles driven has remained fairly constant ~12,000/yr (Source: FHA)
    - Number of vehicles on the road increasing by ~1-2% YOY since 2014 (USDOT)
  - Construction increased from 3.6% to 4.3% of GDP from 2012 to 2017 increasing the amount of oil/lubricant into heavy machinery Source: Dept of Commerce 2017
  - Threat: Electric vehicle sales are up 42% YOY in 2018H1 in Europe; Trend:

Figure 1: Cumulative global EVs sold

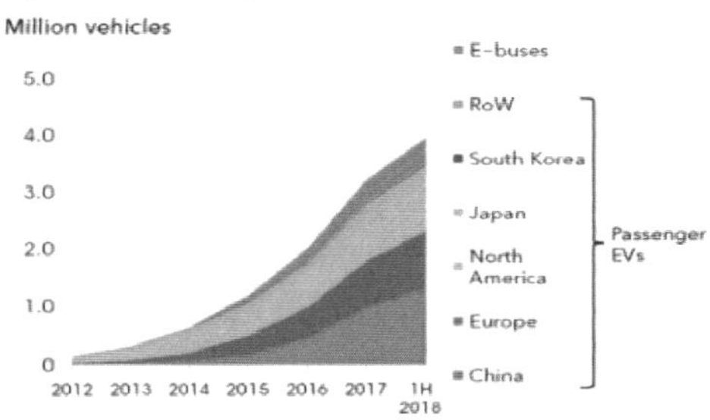

Source: Bloomberg NEF, Marklines

Figure 2: Forecast cumulative global passenger EVs sold

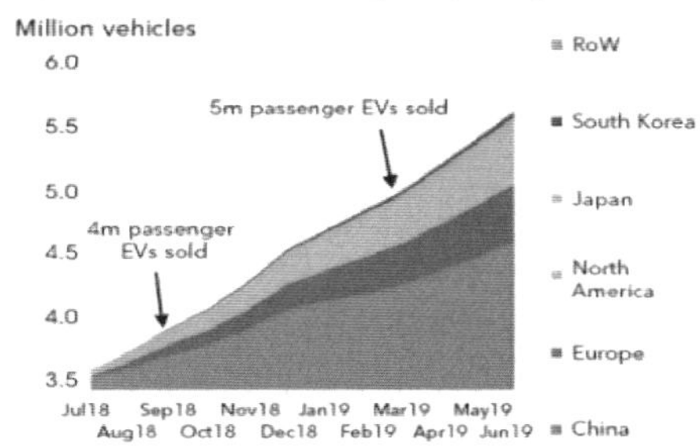

Zochem

Agriculture

- Increased visibility and demand into developing nations through ZNI.

  Zinc sulfate & Zinc oxide (both French Process & American Process)
    - 65% increase of zinc units from 2015 through 2017 (Source: IZA)

  Global population increase from 7.5B to 9.7B by 2050
    - Find more rainfed acreage for cultivation OR improve yield per acre

  Estimated increase from 2017 to 2018 of 35kt (Zn units)

Global Consumption of Zinc Chemicals in Agriculture - 2017

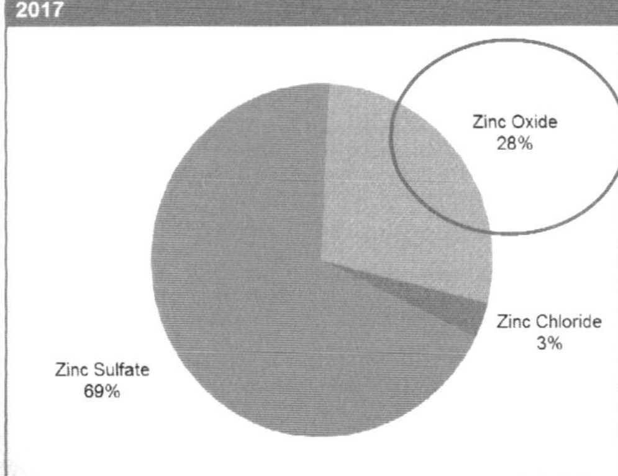

Consumption of Zinc Chemicals in Agriculture by Region - 2017

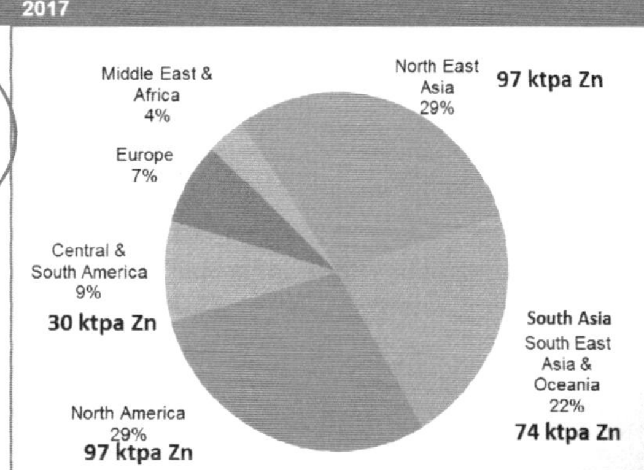

Zochem

Regulatory - ZnO

We, members of the zinc and zinc oxide industries, must do our part to ensure the safe usage and consumption of these materials is well understood. We must lead.

- E.U.
  - Nano ZnO – inhalation route health concern to respiratory system
  - Sunscreen applications – maximum of 25% of nano or bulk ZnO
  - ZnO push over organics environmentally due to speed of organic degradation on the coral reef vs. inorganic solutions
- Human volunteer study (E.U.) completed which is now reporting a new limit value "of concern" to exposure in the workplace. US/Canada/Others at 5 mg/m3; study concerned with levels ~1.5 mg/m3 concerning effects on CRP (inflammation in body)
  - Potential for new workplace exposure limits
- Canada – chemical management plan's risk assessment associated with Zinc and Zinc compounds is complete.
  - Govt of Canada and IZA meeting occurred in Nov 2018. Initial risk assessment has been completed – focus group pre-empting and educating on zinc and zinc compounds to mitigate potential of having additional risk management of zinc compounds for safe handling, transport, labelling, etc. Next Update: May 2019

Source: IZA

Zochem

New/Future Applications

- Continued efforts on:
  - Concrete industry and substitution or working in conjunction with fly ash (retarding the hydration process) and aiding the formability of the concrete. *Source: University of Illinois Urbana-Champaign*
  - Roof shingle substitution of cuprous oxide
    - Cost/benefit analysis
    - Another mature industry
  - Electric/"Tech" applications need to be pushed
    - Semi-conductors (n-type: gives free electrons)
    - Nano-applications in-progress
    - Electric car batteries
  - Reverse substitution on TiO2 in paints/pigments
    - Cost/benefit analysis would drive this potential vs. physical properties. Tough proposition.

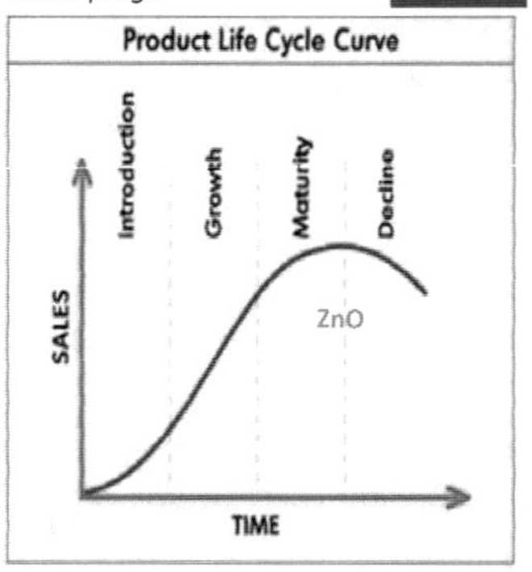

Zochem ZnO

Summary

- Overall zinc oxide market
  - Tires – upward pressure on supplier capacity and meeting new local NA demands from geographical shifts is expected over the next 1-2 years.
    - After market tire replacement considered as 70% of all tires purchased in NA and therefore OEM sales reduction is a data point to monitor but not individual driver
  - Agriculture – fertilizer application growth is 65% for Zn units over last 3 years. $ZnSO_4$ and ZnO both constituents of the increased Zn consumption. Additional participation and education within developing nations has spawned the large increase. We must continue the message of zinc requirement in the daily diet.
  - Specialty Chemical – focus on motor oil industry seeing largest threat for reduced tonnage of ZnO. Slow reduction noted due to increased trucking from NA economy + increased heavy machinery/lubricants from growth and investment
- D.P.E.S.T.O. analysis highlighted:
  - Millennial purchase behaviour
  - Urbanization
  - Increased education on nutrition in developing nations through globalization
  - Electric vehicles/technological changes
  - Political climate (regional + global)

Zochem ZnO

Questions?

Mohit Sharma
msharma@zochem.com
905-453-4100 x4234

Century Zinc Mine Processing Plant

NEW CENTURY
RESOURCES

2019 IZA CONFERENCE PRESENTATION

February 2019

Cautionary Statement

NEW CENTURY
RESOURCES

New Century Resources believes that the production target, forecast financial information derived from that target and other forward looking statements included in this presentation are based on reasonable grounds. However, neither the Company nor any other person, including Sedgman Pty Ltd makes or gives any representation, assurance or guarantee that the production target or expected outcomes reflected in this announcement in relation to the production target will ultimately be achieved.

Investors should note that the Company believes the commodity prices, AUD:USD exchange rate and other variables that have been assumed to estimate the potential revenues, cash flows and other financial information are based on reasonable grounds as at the date of this presentation. However, actual commodity prices, exchange rates and other variables may differ materially over the contemplated mine life and, accordingly, the potential revenue, cash flow figures and other financial information provided in discussions set out in this announcement should be considered as an estimate only that may differ materially from actual results. Accordingly, the Company cautions investors from relying on the forecast information in this announcement and investors should not make any investment decisions based solely on the results.

A number of key steps need to be completed in order to bring the Century Zinc Mine to full scale production. Many of those steps are referred to in this presentation and previously released Restart Feasibility Study announcement. Investors should note that if there are any delays associated with completing those steps, or completion of the steps does not yield the expected results, the revenue and cash flow figures may differ materially from actual results.

To achieve the range of outcomes indicated in this presentation, additional funding in the order of A$63 million will likely be required to achieve full production above the initial A$50 million capital outlay to bring the project into initial production. While the Company has significant cash reserves and anticipated cashflows from operations, investors should note there is no certainty that the Company will be able to raise any additional funding if needed. It is also possible that such funding may only be available on terms that may be dilutive to or otherwise affect the value of the Company's existing shares.

Certain statements contained in this presentation constitute forward looking statements. Forward looking information often relate to statements concerning New Century Resources' future outlook and anticipated events or results and, in some cases can be identified by terminology such as "may", "will", "could", "should", "expect", "plan", "anticipate", "believe", "intend", "estimate", "projects", "predict", "potential", "continue" or other similar expressions concerning matters that are *not* historical facts. Statements of historical fact are not considered forward looking information.

Forward looking statements are based on a number of material factors and assumptions, including, but not limited in any manner to, those disclosed in results; the ability to explore; communications with local stakeholders and community and government relations; status of negotiations of joint ventures; weather conditions; Ore Reserves; Mineral Resources; the development approach and schedule; the receipt of required approvals, titles, licenses and permits; sufficient working capital to develop and operate the mines and implement development plans; access to adequate services and supplies; foreign currency exchange rates; access to capital markets; availability of qualified work force; ability to negotiate, finalise and execute relevant agreements; lack of social opposition to mines or facilities; lack of legal challenges with respect to the property; the timing and amount of future production and ability to meet production, operating and capital cost expenditure targets; timing and ability to produce studies and analysis; execution of the credit facility; ability to draw under the credit facility and satisfy conditions precedent including execution of security and construction documents; economic conditions; availability of sufficient funding; the ultimate ability to mine, process and sell the mineral products produced; the timing, exploration, development, operational, financial, budgetary, economic, legal, social and political factors that may influence future events or operating conditions. Forward looking statement are only predictions based on New Century Resources' current expectations and projections of future events. Actual results may vary from such forward looking information for a variety of reasons.

Forecast financial information provided in this presentation is based on the Restart Feasibility Study. The Company is of the view it has reasonable grounds for providing the forward looking statements included in this presentation. However, the Company cautions that there is no certainty that the forecast financial information derived from the production targets will be realised. The Company confirms that all material assumptions underpinning the production target and forecast financial information contained in the Company's ASX Announcements on 28 November 2017 and 15 January 2018 continue to apply and have not materially changed.

Other than required by law, New Century Resources assumes no obligation to update any forward looking information to reflect, among other things, new information or future events.

COMPANY & PROJECT OVERVIEW

Focused on maximising the value of the Century assets and returns to shareholders

NEW CENTURY RESOURCES

Company Overview

Focused on maximising the value of the Century assets and returns to shareholders

NEW CENTURY RESOURCES

SNAPSHOT

- ASX300 Listed (ASX: NCZ)
- Operations ramping up to being a top 10 zinc producer (+250kt zinc in concentrate per annum)
- +6 years Reserves with mine life upside from insitu Resources (9.3Mt & 10.8% Zn + Pb) & exploration potential
- Targeting first cashflow positive quarter of operations (Q1 2019) and up to A$1.8B in free cash flow from current Reserves

CAPITAL STRUCTURE

Shares on Issue	505M
Unlisted Options (avg. price A$0.42/share*)	114M
Market Capitalisation (at A$0.85/share)	A$430M
Cash & Debt	
cash & receivables (at 15/02/19)	A$32.4M
working cap. facilities^ (20% drawn)	A$100M

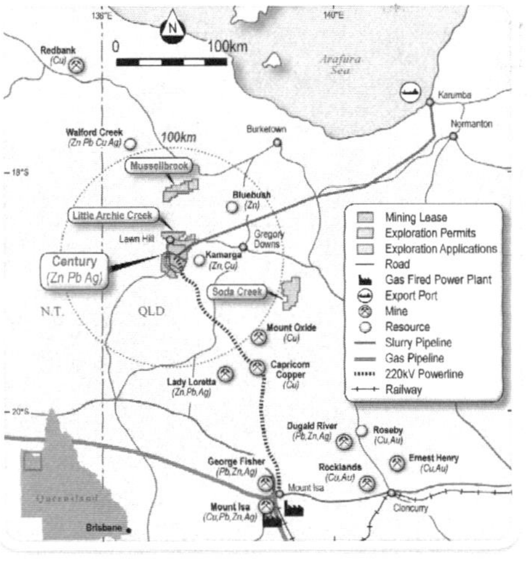

* Options priced from $0.25/share to $1.99/share. Total consideration of A$49.3M if fully exercised

^ A$100M in working capital facilities of which A$40M remains subject to performance hurdles (35 days averaging 45% recovery, >7.3Mtpa mining rate & >8Mpa EBITDA (see ASX ann. 18 Feb 2019)

4

Highlights in the Restarting of One of Australia's Iconic Mines

NEW CENTURY RESOURCES

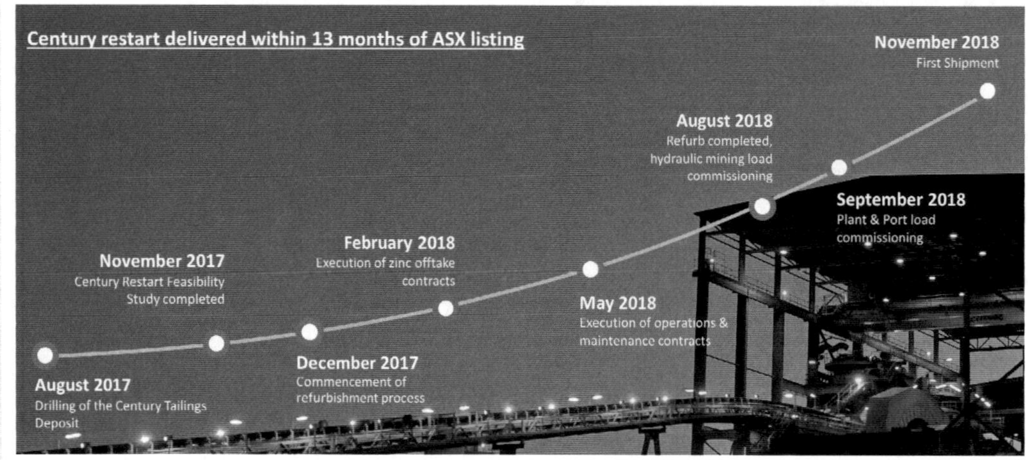

5

Project Overview

Strong value proposition from Century Restart Feasibility Study (completed Dec 2017)

NEW CENTURY RESOURCES

Tailings Ore Reserve & Production Capacity:

- Reserves: 2.3Mt zinc & 29.7Moz silver (77.3Mt at 3.1% ZnEq)
- Production: 264,000tpa zinc & 3Moz pa silver (in 500ktpa of concentrate)
- Targeting ramp up of operations to full capacity by the end of 2019

Mine Life:

- Initial mine life of 6.3 years (tailings Reserves only)
- In-situ Resources to provide mine life extension (9.3Mt at 10.8% Zn+Pb)
- Active exploration programs across 1,400km^2 tenement package

Robust Feasibility Study Economics:

Value Metric (after tax)	At Zinc US$1.25/lb & AUD/USD $0.75 (US$2,750/t or A$3,666/t)
NPV_8	A$1.3 Billion
IRR	270%
Free Cashflow	A$1.8 Billion

6

Forecast Production Profile

Operations ramping up to be a top 10 zinc producer

NEW CENTURY RESOURCES

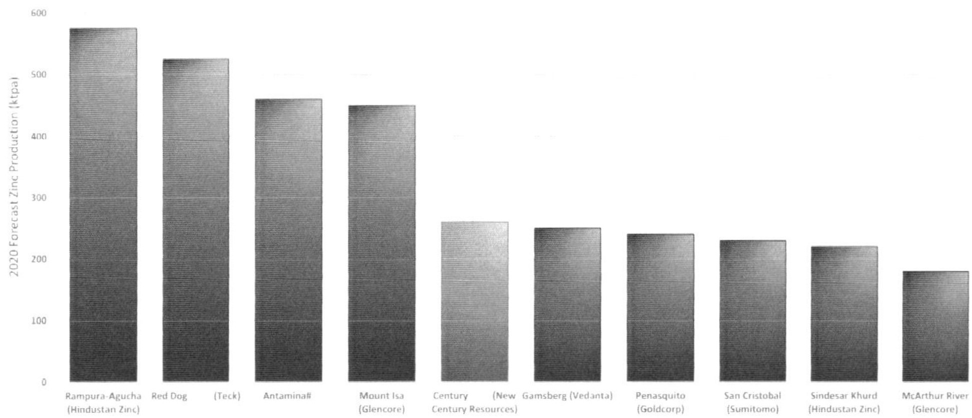

Source: Wood Mackenzie, February 2019 (percentages reflect proportion of forecast 2020 global zinc production)
 # Antamina ownership: BHP Billiton (33.75%), Glencore (33.75%), Teck (22.5%), Mitsubishi (10%)

7

Forecast Cost Profile

Operations forecasted to be in the lowest cost quartile on completion of full ramp up

NEW CENTURY RESOURCES

Restart Feasibility Costs (Dec '17)	A$/t Feed	US$/lb (payable)
Hydraulic Mining	2.75	0.06
Processing Plant	10.31	0.22
Transport, TCs & silver credit	4.63	0.10
C1 Cash Costs	**17.69**	**0.38**
Depreciation	1.48	0.03
C2 Cash Costs	**19.17**	**0.41**
Royalties & Corporate	3.97	0.09
C3 Cash Costs	**23.14**	**0.50**

Zinc Treatment Charges (TCs)

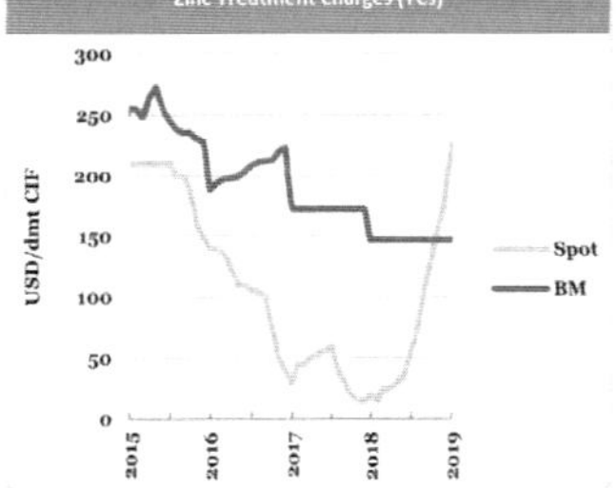

Source: LME 2019 (BM: Annual Benchmark Rate)

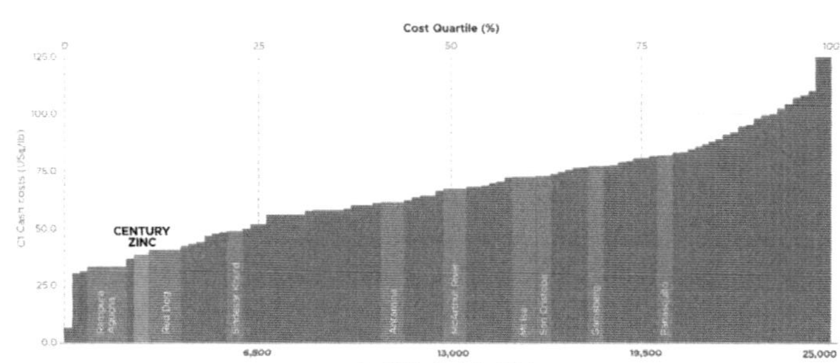

Source: Wood Mackenzie, February 2019 forecasted 2020 C1 cash costs & Century Restart Feasibility C1 costs estimate (Dec '17)

- Century **progressing toward feasibility study opex estimate** in line with continued ramp in overall mining rate & plant recovery
- Recent spot TCs increase (feasibility estimate US$75/t vs current ~US$200/t) implies only a moderate potential increase Century C1 cost projection from US$0.38/lb to US$0.51/lb
- Century remains **on track to be a lowest quartile operation** on completion of full scale ramp up

8

OPERATIONS OVERVIEW

Operations ramping up to be a top 10 zinc producer

NEW CENTURY RESOURCES

Hydraulic Mining operations at the Century Zinc Mine

Operations Underway at Century

Initially undertaking tailings reprocessing to facilitate site rehabilitation

NEW CENTURY RESOURCES

Century Mine Overview

- **Economic Rehabilitation:**
  Effective site rehabilitation generating significant cashflow
- **Tailings/Evaporation Dam Area:** Rehabilitation facilitated by tailings reprocessing operations (35-40% of total rehab)
- **Waste Dumps:**
  A$81M provision for finalising capping of the waste dumps included in cashflow model
- **Full Site Rehabilitation:**
  Facilitated over the life of tailings reprocessing operations

10

Flowsheet of Operations

Identical processing and logistics chain to historical operations

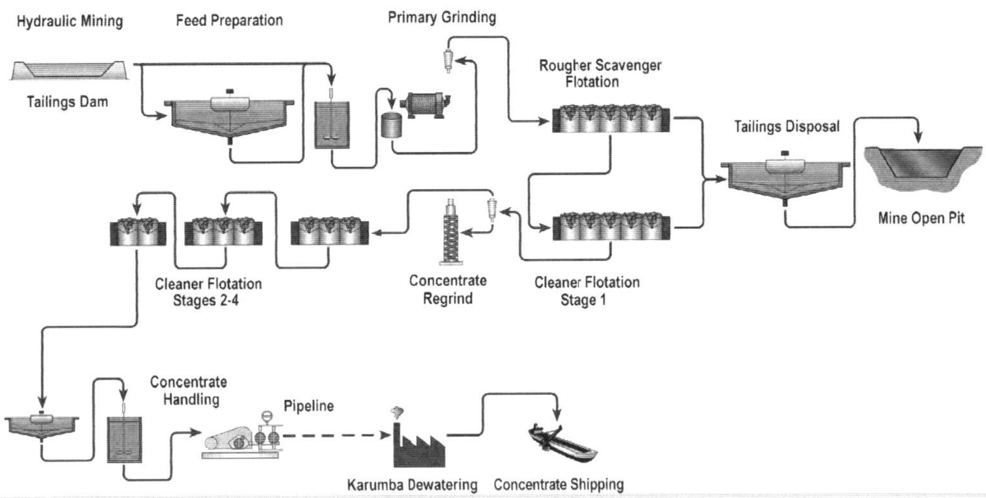

11

Hydraulic Mining Operations

Targeting ramp up to a 15Mtpa mining rate by end of 2019

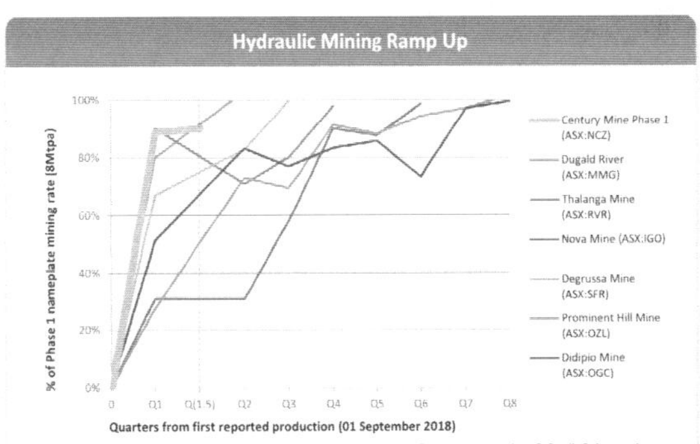

- Hydraulic mining achieving **up to 7.3Mtpa** mining rate
- **Expansion up to 10Mtpa** underway via third mining cannon
- Third cannon on scheduled for end of Q1 2019 and full scale ramp up to 15Mtpa by end of Q4 2019

12

Processing Plant Operations

Plant ramp up continues – currently 75% of overall nameplate zinc recovery target

Century Processing Infrastructure

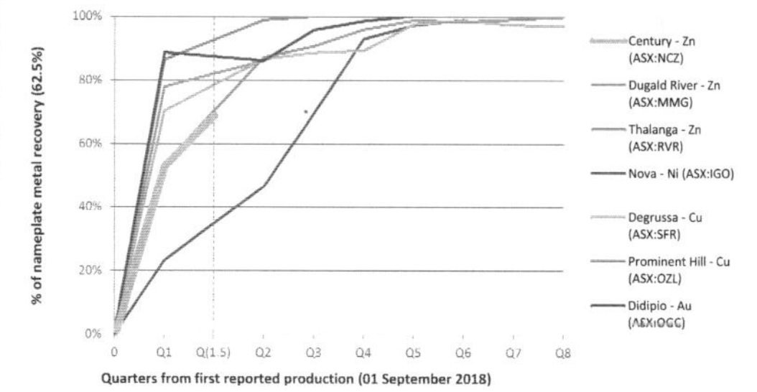

Source: Public company data & Credit Suisse estimates

- Plant now regularly **achieving 45%+** zinc recoveries (target: 60%+)
- Recent results **up to 51% total zinc recovery**
- Continued increase in concentrate grade and decrease in impurities

13

Port & Shipping Operations

Exporting zinc concentrate to smelters in China, Europe and Australia

Karumba Port Facility

Reclaiming of Stockpiled Concentrate

MV Wunma Transhipper

Zinc Concentrate Exports

14

Zinc Concentrate Quality

Continued improvement in zinc concentrate quality during ramp up process

	Century (commissioning grade)	Century (target grade)	McArthur River (Glencore)	Dugald River (MMG)	Gamsberg (Vedanta)	Rampura-Agucha (Hindustan Zinc)	Red Dog (Teck)
Zn	48 - 51 %	51 - 54.5 %	47%	52%	50%	51.5%	55.6%
Fe	1.5 - 3.0 %	0.8 - 2.0 %	5.9%	11.0%	8.9%	8.8%	5.0%
Mn	<0.15 %	<0.15 %	<0.01 %	2.0%	2.6%	0.3%	<0.01 %
SiO_2	5.0 - 7.5 %	5.0 - 7.5 %	4.6%	3.5%	2.0%	3.3%	4.5%
C	3.0 - 4.5 %	3.0 - 5.0 %	<0.1%	<0.1%	0.2%	0.9%	<0.1%
Cd	0.08 - 0.15 %	0.08 - 0.15 %	0.12%	<0.1 %	<0.1 %	0.18%	0.4%
As	<0.01 %	<0.01 %	0.20%	0.02%	<0.01 %	0.01%	0.02%
Hg	<15 ppm	<50 ppm	40 ppm	15 ppm	22 ppm	43 ppm	80 ppm
Cu	<0.6 %	<0.6 %	1.1%	0.2%	0.1%	0.1%	0.2%
Pb	4.0 - 6.0%	1.2 - 3.0 %	4.6%	0.2%	0.1%	1.0%	3.8%
S	27 - 30 %	27 - 30 %	30.0%	31.0%	29.4%	31.0%	31.7%
Ag	130 - 200 ppm	50 - 250 ppm	130 ppm	80 ppm	2 ppm	91 ppm	150 ppm

Source: Wood Mackenzie, August 2018 (peer concentrate specifications are indicative only and may not represent current concentrate qualities)

- **Significant reduction in lead content** of 'commissioning grade' concentrate, in line with increasing overall zinc recoveries
- Century zinc concentrate remains in strong demand for its **low iron and manganese impurity** levels
- **80,000t of production sold to date** (spot contracts) and first sales under long term offtakes are underway

MINE LIFE EXPANSION

Strong potential for mine life extension through existing Mineral Resources and exploration

NEW CENTURY RESOURCES

Underground 3D visualisation of the South Block Deposit

Century Insitu Mineral Resources

Opportunity for mine life extension – PFS nearing completion

NEW CENTURY RESOURCES

17

Century Exploration Potential

Identified drill targets to be tested post wet season (Q2 2019)

NEW CENTURY RESOURCES

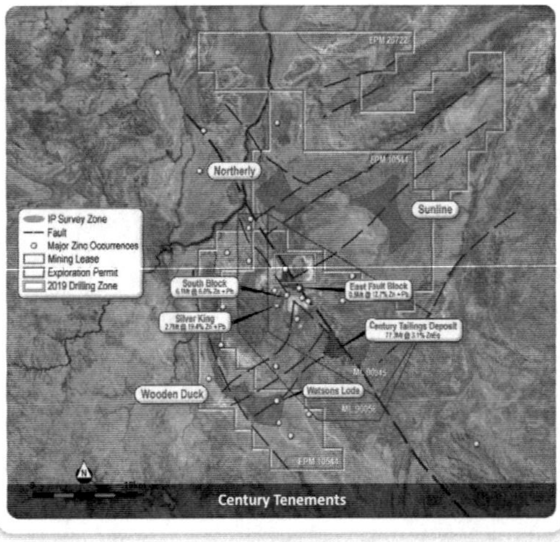

Exploration Focus & Context:

- Original Big Zinc ore body cut off in the north face by faulting
- Opportunity to discover missing 'displaced' section of the original ore body

2019 Drilling Plans:

- Targeted IP program directly north of the original Century open pit has identified drill ready opportunities
- Targets located on the mining lease (opportunity for near term development on discovery)
- Proposed drilling of identified IP targets adjacent to the open pit to occur on completion of wet season (Q2 2019)

18

JORC 2012 Compliant Reserves & Resources Statement

NEW CENTURY RESOURCES

Mineral Resources	Tonnes (Mt)	Grade			Contained Metal		
		Zinc (%)	Lead (%)	Silver (g/t)	Zinc (t)	Lead (t)	Silver (oz)
South Block (Indicated)	6.1	5.3	1.5	43	322,000	90,000	8,550,000
Silver King (Inferred)	2.7	6.9	12.5	120	186,000	337,500	10,500,000
East Fault Block (Inferred)	0.5	11.6	1.1	48	60,000	5,500	800,000
Total Mineral Resources	**9.3**	**6.1**	**4.7**	**66**	**568,000**	**433,000**	**19,850,000**
Ore Reserves	**Tonnes (Mt)**	**Grade**			**Contained Metal**		
		ZnEq (%)	**Zinc (%)**	**Silver (g/t)**	**Zinc (t)**	**Lead (t)**	**Silver (oz)**
Century Tailings (Proved Ore Reserve)	**77.3**	**3.1**	**3.0**	**12**	**2,287,662**	**-**	**29,734,819**

Competent Person Statement & ZnEq Calculation

ZnEq was calculated for each block of the Century Tailings Deposit from the estimated block grades. The ZnEq calculation takes into account, recoveries, payability (including transport and refining charges) and metal prices in generating a zinc equivalent value for each block grade for Ag and Zn. ZnEq = Zn%+ + Ag troy oz/t*0.002573. Metal prices used in the calculation are: Zn US$3,000/t, and Ag US$17.50/troy oz.

The information in this announcement that relates to the Mineral Resources estimate on the Silver King Deposit and the East Fault Block Deposit was first reported by the Company in its prospectus released to ASX on 20 June 2017, and the South Block Deposit was first reported by the Company to the ASX on 15 January 2018. The Company confirms that it is not aware of any new information or data that materially affects the information included in the original market announcements and that all material assumptions and technical parameters underpinning the Mineral Resources estimates in the relevant original market announcements continue to apply and have not materially changed.

The information in this announcement that relates to the Ore Reserve estimate at the Century Tailings Deposit was first reported by the Company in its ASX announcement titled "New Century Reports Outstanding Feasibility Results that Confirm a Highly Profitable, Large Scale Production and Low Cost Operation for the Century Mine Restart" dated 28 November 2017. The Company confirms that it is not aware of any new information or data that materially affects the information included in the original market announcement and that all material assumptions and technical parameters underpinning the estimates in the original market announcement continue to apply and have not materially changed.

19

Contact

Patrick Walta

Managing Director

info@newcenturyresources.com

New Century Resources Limited
Level 4, 360 Collins Street, Melbourne VIC 3000
+61 (3) 9070 3300
www.newcenturyresources.com

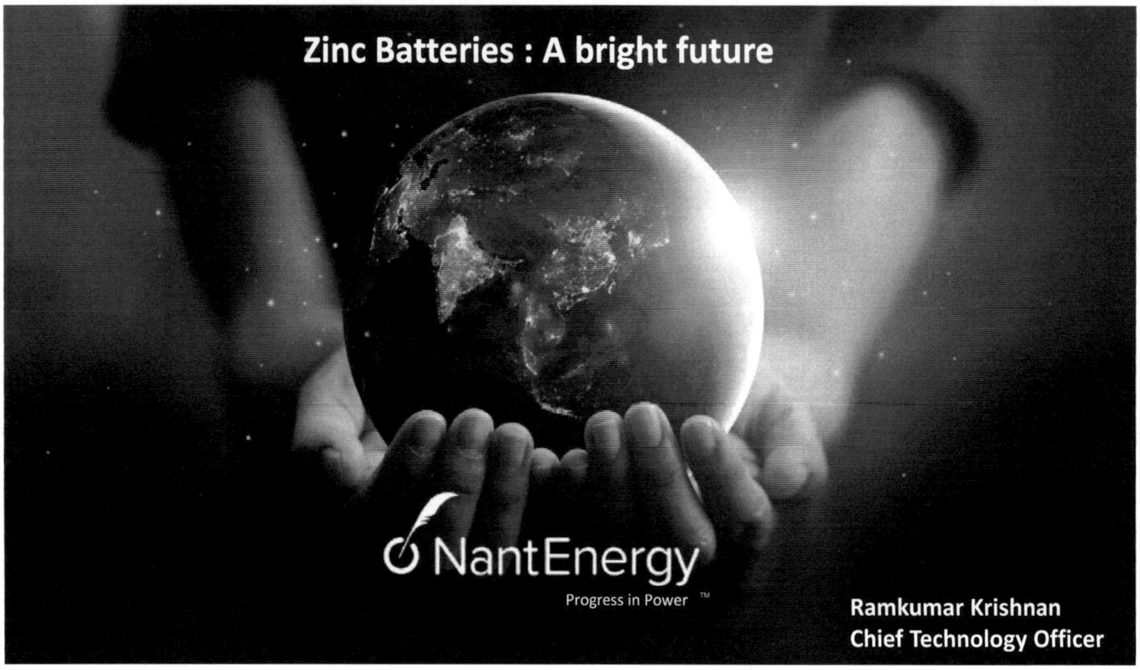

NantEnergy

Technology Disruption & Exponential Growth

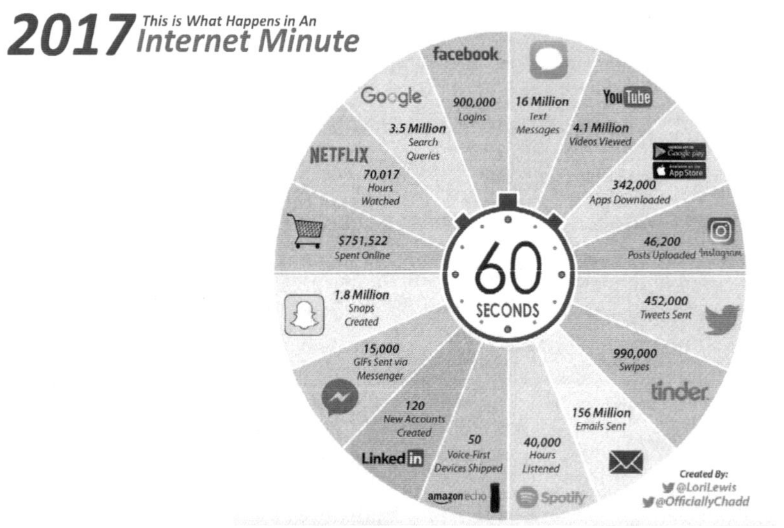

NantEnergy

Technology Disruption & Exponential Growth

2018 *This is What Happens in An Energy Hour*

Global Electricity Production 3000GWh in 1 hour

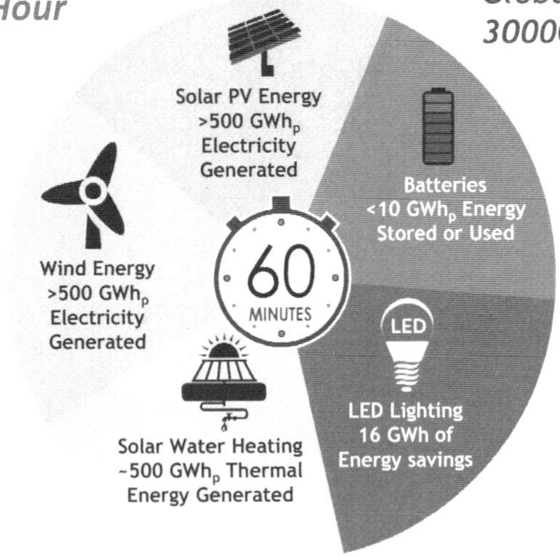

NantEnergy

Technology Disruption - Energy

Technologies that dominated energy generation & storage for over 100 years

Diesel Engine 1893

Lead Acid Battery 1859

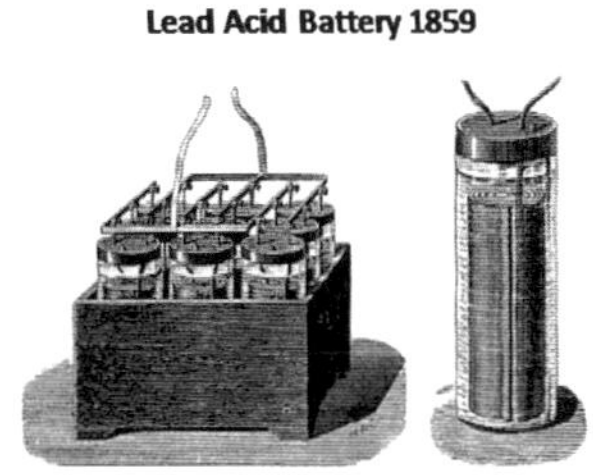

Diesel and Gas Generators in datacenters and utility power plants 2017

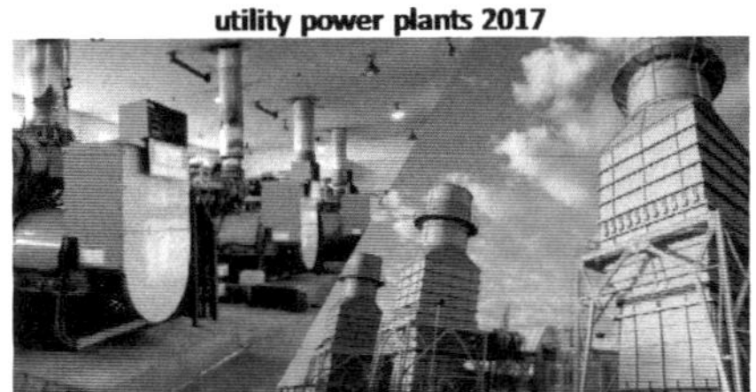

ICE

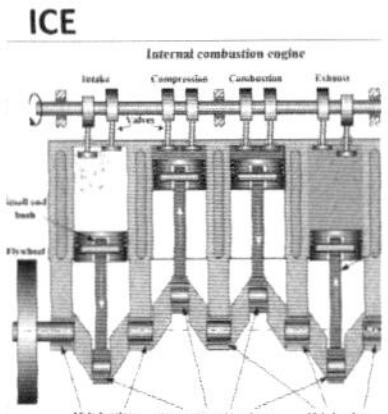

NantEnergy

Technology Disruption - Energy Storage

Stand-alone DG-less Microgrids with Solar and NantEnergy Storage

Source: NantEnergy (Fluidic Energy)

Electric Vehicles

Source: Tesla

The global power sector is about to be fundamentally transformed

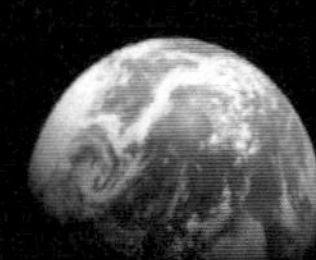

What are the tipping points?

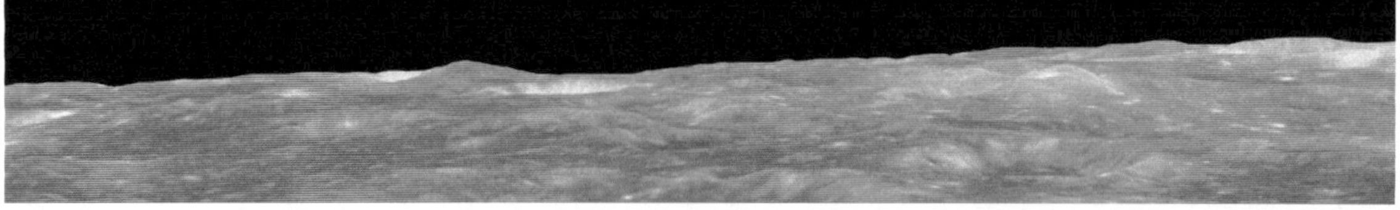

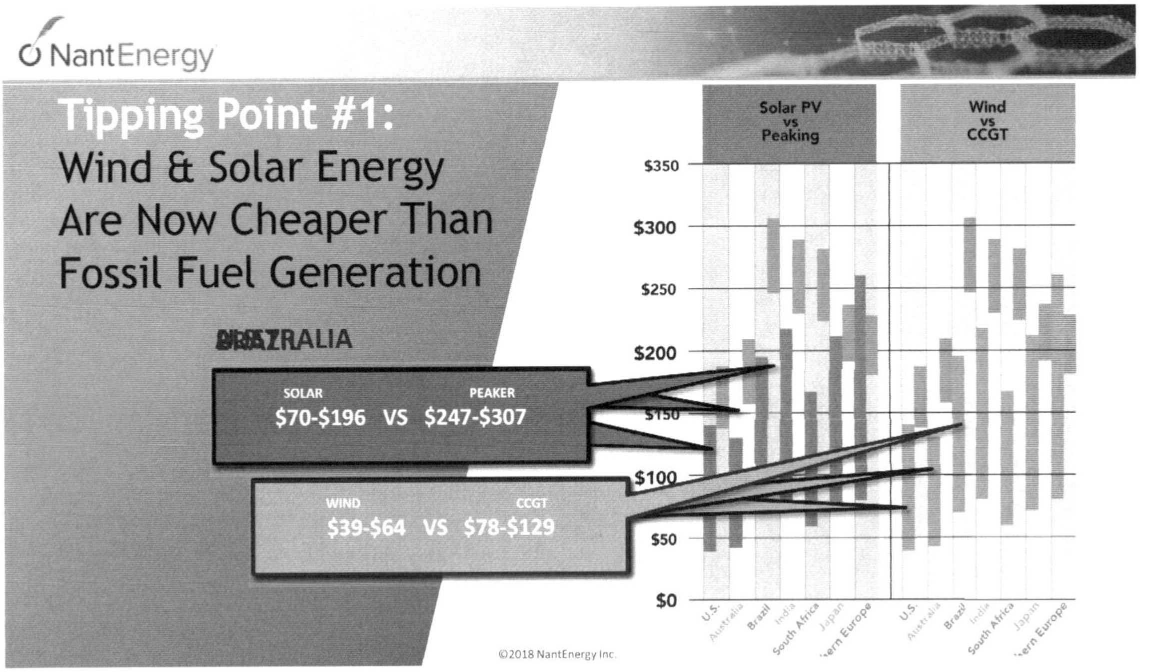
NantEnergy
Tipping Point #1:
Wind & Solar Energy Are Now Cheaper Than Fossil Fuel Generation
SOLAR PEAKER
$70-$196 VS $247-$307
WIND CCGT
$39-$64 VS $78-$129
Solar PV vs Peaking
Wind vs CCGT
$350
$300
$250
$200
$150
$100
$50
$0
U.S.
Australia
Brazil
India
South Africa
Japan
U.S.
Australia
Brazil
India
South Africa
Japan
©2018 NantEnergy Inc.

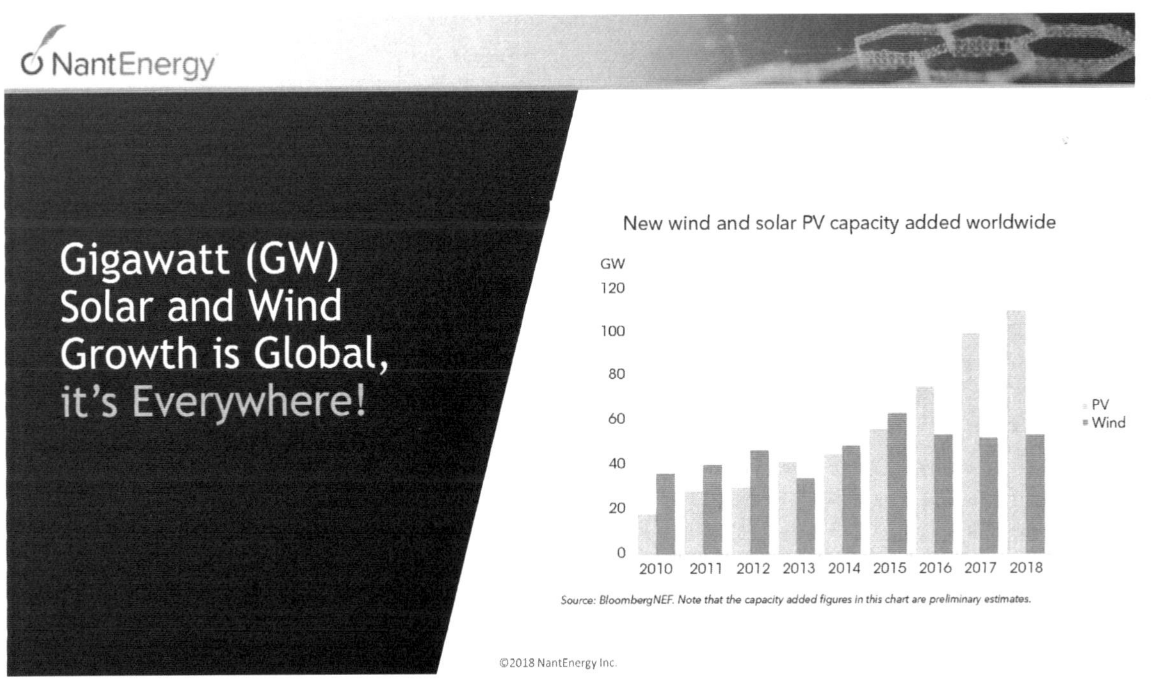
NantEnergy
Gigawatt (GW) Solar and Wind Growth is Global, it's Everywhere!
New wind and solar PV capacity added worldwide
GW
120
100
80
60
40
20
0
2010 2011 2012 2013 2014 2015 2016 2017 2018
PV
Wind
Source: BloombergNEF. Note that the capacity added figures in this chart are preliminary estimates.
©2018 NantEnergy Inc.

NantEnergy

Low cost energy storage is needed to enable full utilization and realization of renewables potential

Renewable-energy powered grids no longer need baseload generation

Source: California Energy Storage Alliance CAISO OASIS data (2016) CEC System Almanac

Source: CAISO OASIS data (2016), 2030 IRP Proposed Reference System Plan Scenario

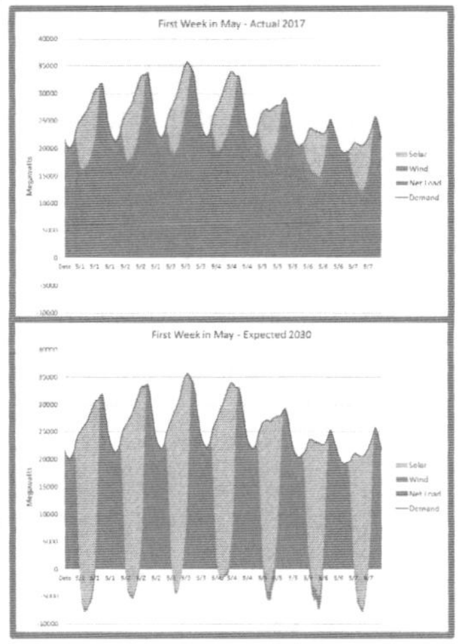

CA Today (29% renewable)

Net load is met by Flexible Gas, Baseload Gas, Nuclear, Geothermal, Imports/Exports, and Curtailments

CA in 2030 (57% renewable)

Net load will need to be met by a combination of Flexible Resources, Imports/Exports, and Curtailment

NantEnergy

Low cost energy storage will enable renewable microgrids: Especially for increasing frequency of billion-dollar disasters

NantEnergy Microgrid in North Carolina

Energy Storage Cabinets and Solar PV Supported through Multiple Hurricanes

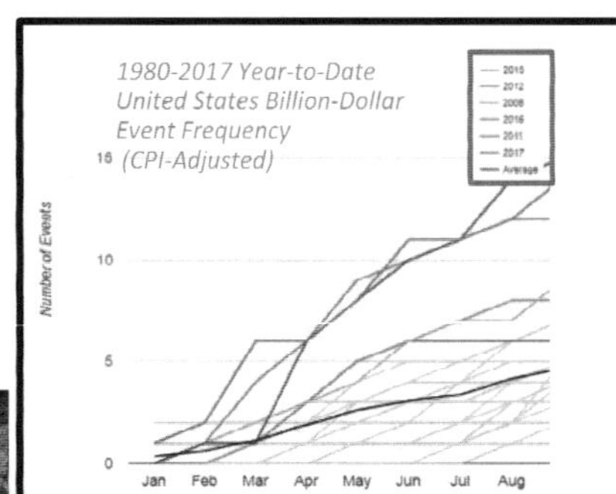

Source: NOAA Climate.gov: https://www.climate.gov/news-features/blogs/beyond-data/2017-us-billion-dollar-weather-and-climate-disasters-historic-year January 2018

NantEnergy

NantEnergy

Low cost energy storage will enable energy access for ~1 billion people globally

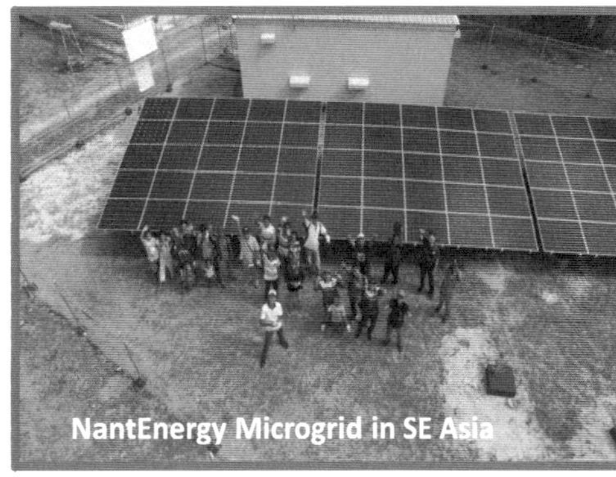

Energy Storage Cabinets and Solar PV

NantEnergy

Tipping Point #2:

Energy Storage Costs are falling and significant growth in market

NantEnergy's Rechargeable Zinc Air Battery Breaks the $100 kWh Cost Barrier ... TODAY

Zinc-Air Breaks the $100 kWh Cost Barrier in 2018

Li-ion Projected to Break the $100 kWh Barrier by 2025

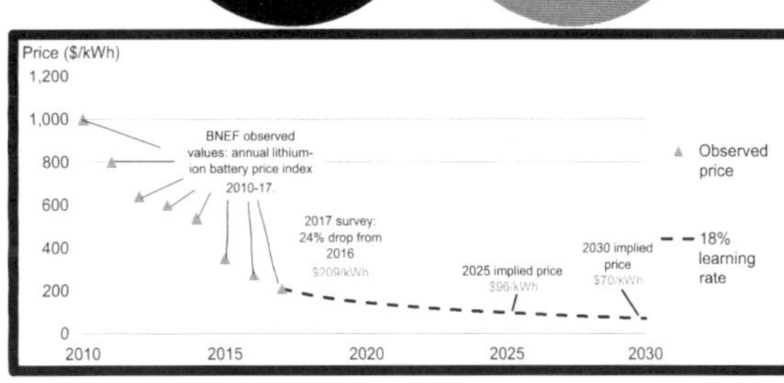

NantEnergy

Market Analysts Have Identified Energy Storage as the Next Major Growth Trend

IHS Market predicts an additional 56 GW of cumulative storage installed between now and 2025.

Bloomberg New Energy Finance's forecasts are directionally similar

How big can the energy storage market be?

2017	2025
4.6 GW Cumulative installed base	56 GW Cumulative installed base
7.7 GWh Cumulative installed base	141 GWh Cumulative installed base
\$1.9 Bn Annual hardware revenues	\$7.9 Bn Annual hardware revenues
0 Markets > 1 GW	9 Markets > 1 GW

NantEnergy

Tipping Point #3: NantEnergy can globally scale using safe, earth abundant local materials and local manufacturing

Example: Manufacturing Plant in Indonesia (2017-2018)

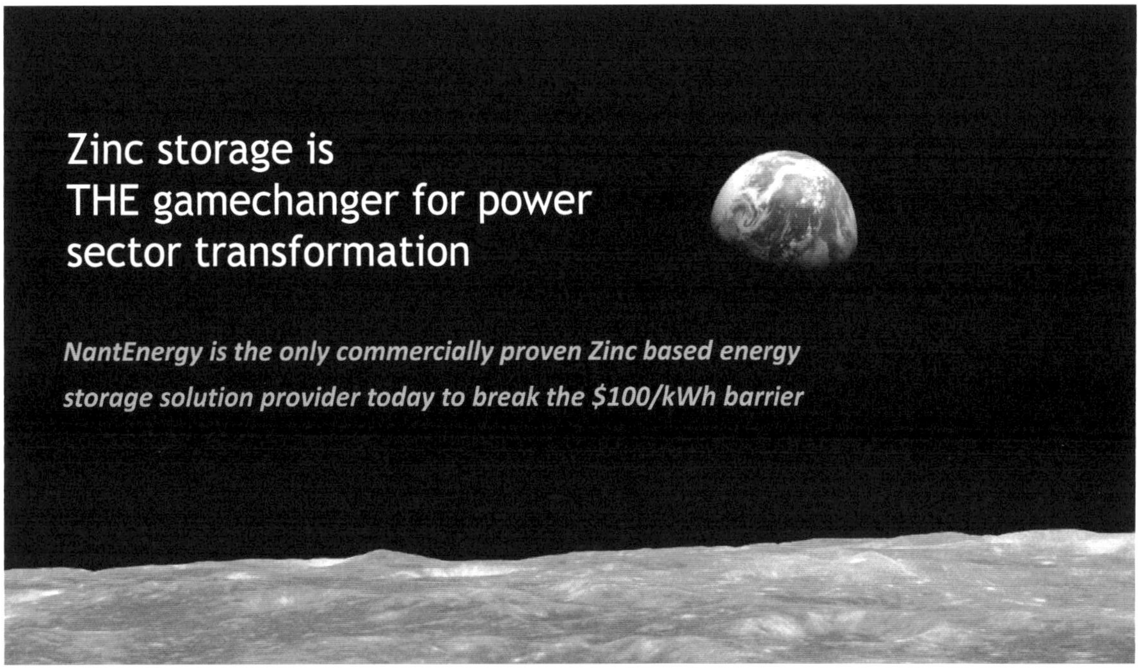

NantEnergy

About NantEnergy

- NantEnergy was founded in 2006, our proprietary advanced rechargeable Zinc- Air battery technology provides low cost, dependable, and clean alternatives to diesel gensets and lead-acid batteries
- More than $200M has been invested to date, which is protected by more than 120 patents
- NantEnergy delivers targeted applications for microgrids, telecom, and commercial and industrial businesses
- We've installed more than 3,000 systems (including 120,000 batteries) in 10 countries; 6.3 gigawatt hours of electricity dispatched in 2018
- Headquartered in Los Angeles & Phoenix, NantEnergy currently has operations in the U.S. and a Licensed Regional Manufacturer in Indonesia covering the ASEAN region
- Subsidiary of NantWorks, backed by renowned billionaire technologist Dr. Patrick Soon- Shiong
- Fall 2018, NantEnergy acquired energy storage division of Sharp Electronics

IFC International Finance Corporation WORLD BANK GROUP

12th Annual Energy Innovation Pioneers 2017

arpa-e

Global Cleantech 100

New Energy Pioneers 2017 Bloomberg WINNER

Green Innovation + Finance Awards Fast-Growth Champion of the year WINNER 2017

NantEnergy

Global Zinc Production and Reserves

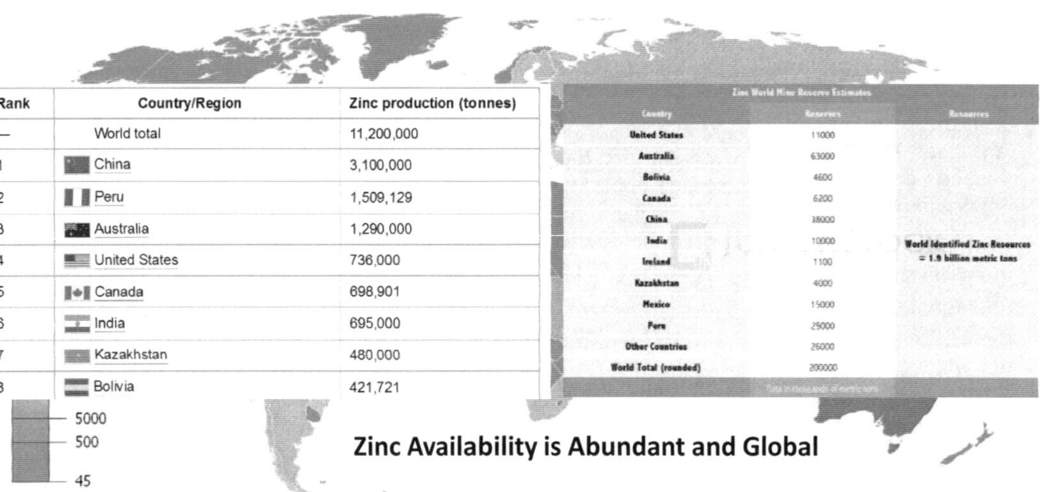

Rank	Country/Region	Zinc production (tonnes)
—	World total	11,200,000
1	China	3,100,000
2	Peru	1,509,129
3	Australia	1,290,000
4	United States	736,000
5	Canada	698,901
6	India	695,000
7	Kazakhstan	480,000
8	Bolivia	421,721

Zinc World Mine Reserve Estimates

Country	Reserves	Resources
United States	11000	
Australia	63000	
Bolivia	4600	
Canada	6200	
China	38000	
India	10000	World Identified Zinc Resources = 1.9 billion metric tons
Ireland	1100	
Kazakhstan	4000	
Mexico	15000	
Peru	25000	
Other Countries	26000	
World Total (rounded)	200000	

Zinc Availability is Abundant and Global

NantEnergy

NantEnergy

Zinc Batteries

Zinc Battery Chemistries

- Zinc Manganese Oxide
- Zinc Bromine
- Zinc Air
- Nickel Zinc
- Zinc Silver
- Zinc ion

Primary Zinc Battery

Alkaline battery Zn/MnO_2

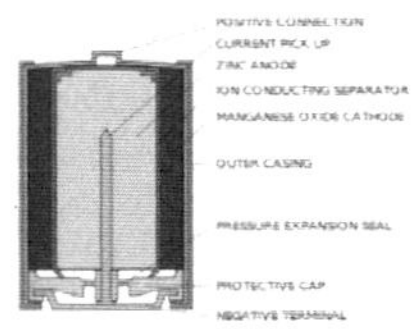

The half-reactions are:

$Zn_{(s)} + 2OH^-_{(aq)} \rightarrow ZnO_{(s)} + H_2O_{(l)} + 2e^-$ $[E_{oxidation}{}^{\circ} = +1.28\ V]$

$2MnO_{2(s)} + H_2O_{(l)} + 2e^- \rightarrow Mn_2O_{3(s)} + 2OH^-_{(aq)}$ $[E_{reduction}{}^{\circ} = +0.15\ V]$

Overall reaction:

$Zn_{(s)} + 2MnO_{2(s)} \rightleftharpoons ZnO_{(s)} + Mn_2O_{3(s)}$ $[e^{\circ} = +1.43\ V]$

❑ The primary Alkaline batteries (over 10 billion individual units produced worldwide) account for

- 80% of manufactured batteries (US)
- 46% of all primary battery sales (Japan)
- 68% (Switzerland), 60% (UK), and 47% (EU) of all battery sales including secondary types.

NantEnergy

About Zinc-Air Technology

Metal-Air is at Nexus of Best Storage Attributes

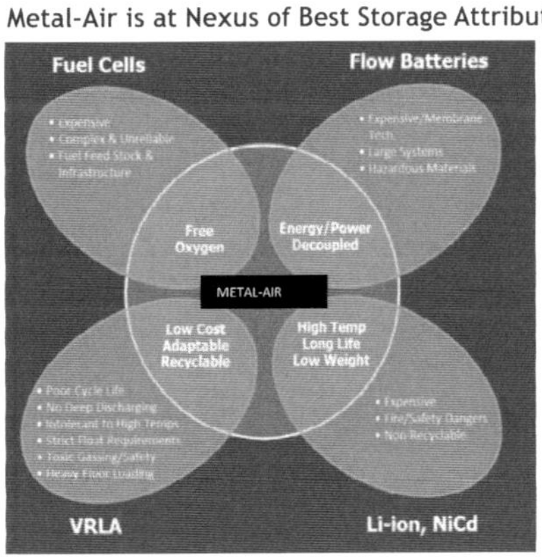

Rechargeable Zn-Air Technology

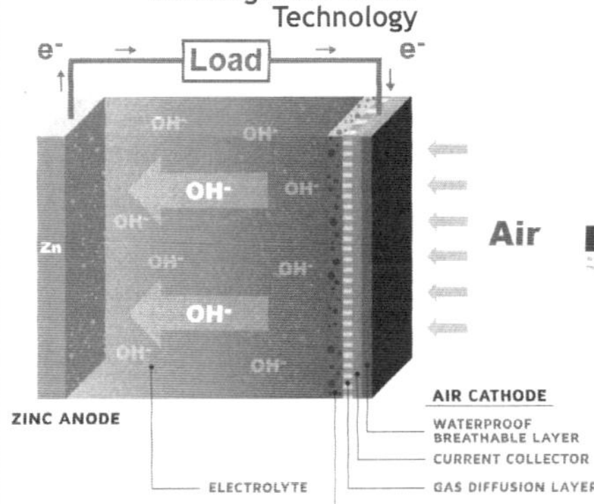

Hybrid Energy Storage

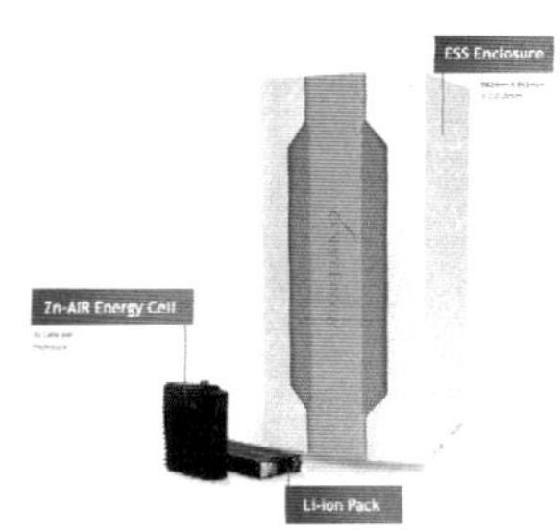

NantEnergy

Metal-Air Chemistries

Table 1
Parameters and reactions of various metal–air batteries.

X. Zhang, et al., Recent progress in rechargeable alkali metal air batteries, Green Energy & Environment (2016)

Batteries	Voltage (V)	Theoretical specific capacity (Ah/kg)	Theoretical energy density (Wh kg^{-1})	Reaction
Al–air	2.71	1030	2791	$4Al + 3O_2 + 6H_2O \rightarrow 4Al(OH)_3$
Mg–air	3.09	920	2843	$Mg + 1/2O_2 + H_2O \rightarrow Mg(OH)_2$
Zn–air	1.65	658	1085	$Zn + 1/2O_2 \leftrightarrow ZnO$
Li–air	2.96	1170	3463	$2Li + O_2 \leftrightarrow Li_2O_2$
Na–air	2.27	487	1105	$Na + O_2 \leftrightarrow NaO_2$
	2.33	687	1600	$2Na + O_2 \leftrightarrow Na_2O_2$
K–air	2.48	377	935	$K + O_2 \leftrightarrow KO_2$

Metal-Air	Benefits	Limitations
Al-air	High theoretical energy density, high cell voltage, inexpensive, abundant metal reserves	High corrosion rates in acidic and alkaline medium, requires non-aqueous chemistry for rechargeability, air electrode stability
Li-air, Na-Air, K-air, Mg-Air	High theoretical energy density, high cell voltage	Dendrite formation, flammable, toxic and non-aqueous chemistries, side reactions, air electrode stability
Zn-air	Inexpensive, stable aqueous chemistry, low corrosion, safe, non-toxic, recyclable, non-flammable, abundant metal reserves	Dendrite formation, air electrode stability
Fe-air	Inexpensive, stable aqueous chemistry, safe, non-toxic, recyclable, non-flammable, abundant metal reserves	Low cell voltage, high corrosion rates in acidic/alkaline medium, air electrode stability

NantEnergy

Zinc Air Breakthroughs in Four Key Areas

Protected by strong IP (120+ patents), trade secrets and technology head start of over 5 years

Mechanical

Electrode Architecture

Chemistry and Electrode Materials

Intelligent Controls

NantEnergy

NantEnergy

Zinc Energy Storage Technology Benefits

NantEnergy Disrupts with Lower Cost, Higher Performance, and Scalable, Modular Platform with Remote Monitoring & Control Advantages

Lowest Cost

- 1/4th raw material cost of lead-acid and 1/17th Li-ion 1/20th cost to scale production capacity
- Only battery that can beat diesel genset costs

Longer Life & Safer

- 3+ times longer life than comparable lead-acid batteries
- No thermal runaway concerns and environmentally friendly

Built-in Intelligence (IoT)

- First battery with fully integrated intelligence
- Provides valuable business intelligence and battery control

Low Supply Chain Risk

- Manufactured using abundant, non toxic local materials
- No precious or rare metals

Battery Technology	Electrode Raw Material Cost ($/kWh)	Global Reserves of Raw Material (MT)	Annual Mining Production (MT)	Top Countries Producing Raw Material
NantEnergy Zinc-Air	1.90	220	13.00	China, Australia, Peru, USA
Primary Zinc	7.10	220	13.00	China, Australia, Peru, USA
Lead Acid	8.00	89	5.00	China, Australia, USA, Peru
Nickel-Zinc	17.90	79	2.53	Philippines, Russia, Australia, Canada
Lithium-ion	35.20	6-30	0.03	Chile, Argentina, Australia
Nickel MH	36.50	62	1.50	Philippines, Russia, Indonesia

NantEnergy

NantEnergy

Zinc-Air Energy Storage System

Complete Modular Outdoor System in IP55 Cabinet

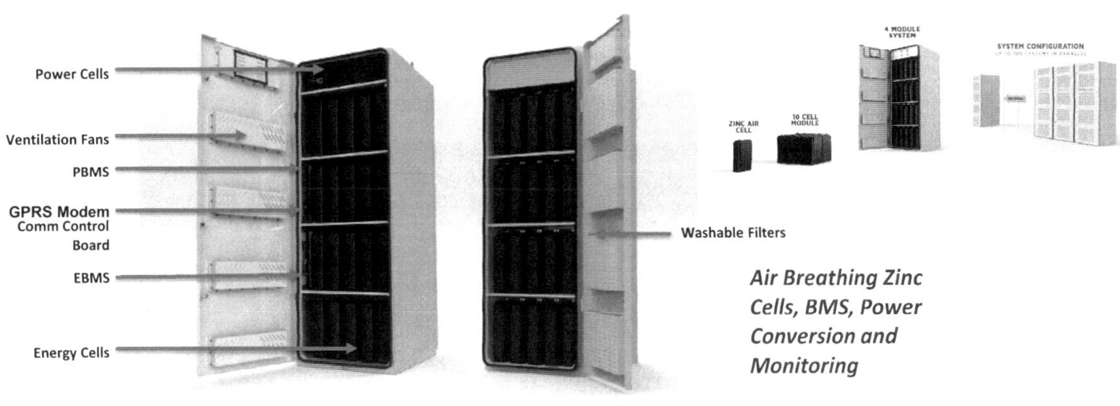

Air Breathing Zinc Cells, BMS, Power Conversion and Monitoring

NantEnergy

NantEnergy Novel Hybrid ZnAir/Li-ion ESS that is Both Energy Efficient & Long Duration

Hundreds of Hybrid Systems Deployed in Asia, Africa and Central/Latin America

Sprinter Storage Li-ion

- Fast charge capability
- High cycle life
- High energy efficiency
- Up to 2 Hours

Marathon Storage Zinc Air

- Long calendar life
- Low initial cost
- High cycle life
- Ability to endure 100%, 0%, and partial SOC
- Low self discharge
- Low standby/cooling losses
- Up to 48 Hours

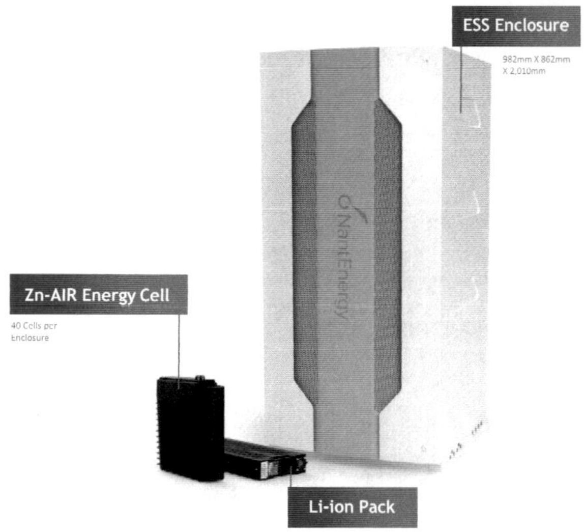

NantEnergy

NantEnergy Zinc Air Breathing Cell Batteries Operating Worldwide

- Performance validated by repeat orders and no service level agreement penalties since initial installations in 2011
- 3,000+ systems (40 cells) operating outdoors (with no cooling) for large commercial customers in 10 countries
- Built-in IoT intelligence – advanced monitoring, data capture and analytics
- 100% Communicating 24/7 to NantEnergy HQ for over 6 years
- Leader in storage for PV microgrids, 100+ installed (largest 1.6MWh) with no diesel gensets needed
- Ultra-low capital cost to manufacture

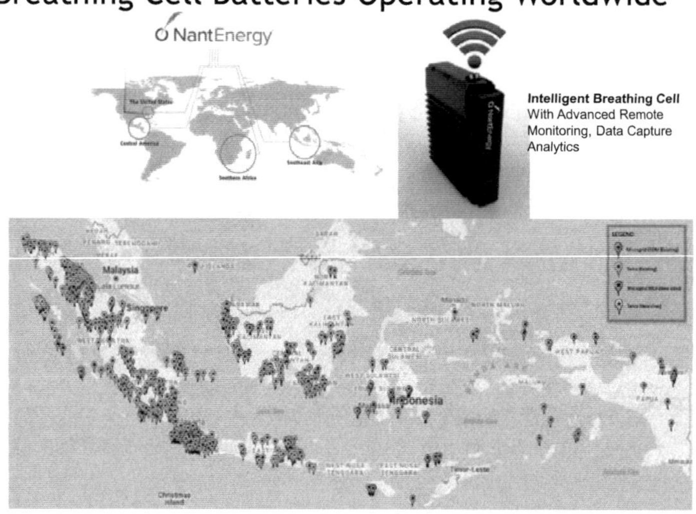

Selected View of Micro-grid Remote Monitoring of NantEnergy Systems in SE Asia

NantEnergy

NantEnergy: Harnessing Nature & Technology to Solve Carbon Pollution

Air Breathing Zinc Rechargeable Battery: Smart Storage System™
The World's First Globally Scaled Deployed Low-Cost Green Renewable Energy Storage System

Global Deployment	Proven Resilience	Human Impact	Environmental Impact	Intelligent System	Sustainable Cost
Over 1,000+ Deployed Sites in 9 Countries	Average 2,000 Cycles Per Cell >1,000,000 Cycles Across all Cells	Breathing Cell Providing Sole Source Power to 200,000 People to Date	Over +10,000 Metric Tons of CO_2 Reduction in 2017 Alone	24/7 x 365 Days Communication	$100 Broke the Barrier of $100 per kWh in 2018
Deployed in Over 110 Villages in Africa and Asia	Longest Cell Cycled: >30,000 Cycles >5,000 Discharge Hrs >5 Years Deployed	Telecom Towers Covering 4,000,000+ People	Prevented the Discharge of Over +50,000 Metric Tons of CO_2 To Date	Globally Controlled Network Operations	Automated Manufacturing
Deployed in Over 1,000 Cell Tower Sites Worldwide	Over 55MWh Total Capacity Installed >120,000 Cells	Eliminated 1 Million Lead Acid Batteries	No Lead Materials in Battery Systems	Intelligent Battery Management System	Enable Local Manufacturing
+3,000 Systems of 40 Cells Per System Installed Globally	Over 6 Years of Real World Energy Storage	Eliminated 4 Million Liters of Diesel Fuel	No Lithium or Cobalt Reducing Risk of Fire or Explosions	All Battery Systems Connected via Cellular Link	Readily Available Plentiful Materials & Parts

NantEnergy

NantEnergy

Zinc Air Stationary Energy Storage Markets

Telecom

Microgrid

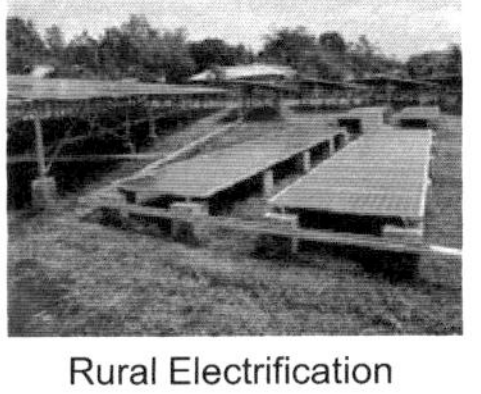
Rural Electrification

Data Center Backup

Critical Power

Critical Utility Applications, SCADA, Communications

Commercial Emergency

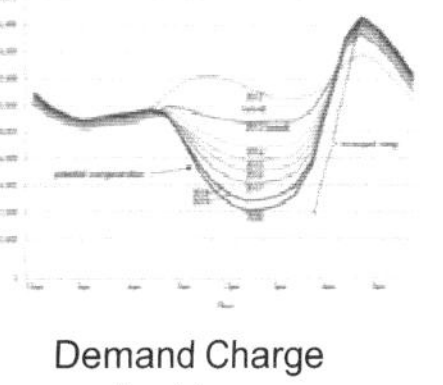
Demand Charge Avoidance

14

NantEnergy

NantEnergy

Zinc-Air in Telecom Applications

Proven product with repeat orders over 6 years

- 6 years of proven field performance, 1.2+ Million cycles cumulative, 9 countries
- Robust performance even in the toughest field conditions - Batteries have even handled gun shots and battery chemistry is proven to be safe
- Elimination of ground water pollution due to Lead in remote regions.
- >30% TCO savings for customers
- Large deployments (~1400 sites) with several MNO's
- Successful completion of every customer demo at largest and most demanding Telecom companies
- Double digit market share with one of the largest Telecom company in the world
- Major MNO minimizing Lead-acid and Genset purchases and theft throughout their region

NantEnergy

Demand Management and Grid Resilience

5 million

- Number of US utility customers pay $15-20/KW demand charge
- Demand charges can make up more than 50% of power bill in some cases

3526

- Number of power outages during 2017
- 438 just in California with average outage of ~ 1hour
- 1 hour of downtime costs a company average of ~$100K of lost revenue

36.7 million

- Number of people in the US affected by power outages

Source: Eaton Blackout tracker Annual report

NantEnergy

combined with Smart Energy Management

Built-in Intelligence

Unique Customer Benefits

World Class Reliability Via
Redundancy & Diagnostics

100% Cloud
Connected 24/7

Whole Site Asset
Management

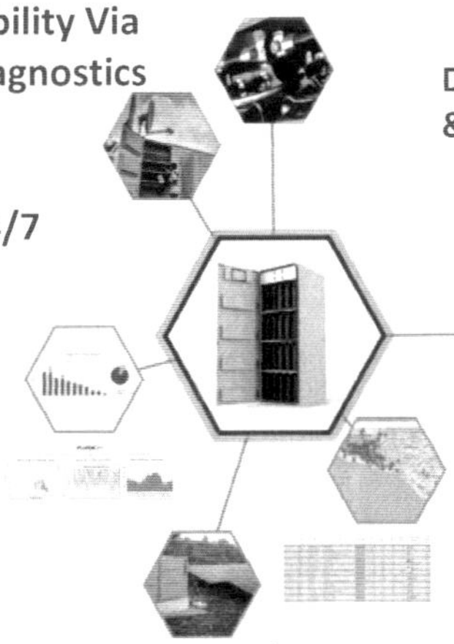

Demand Management
& Grid Resilience

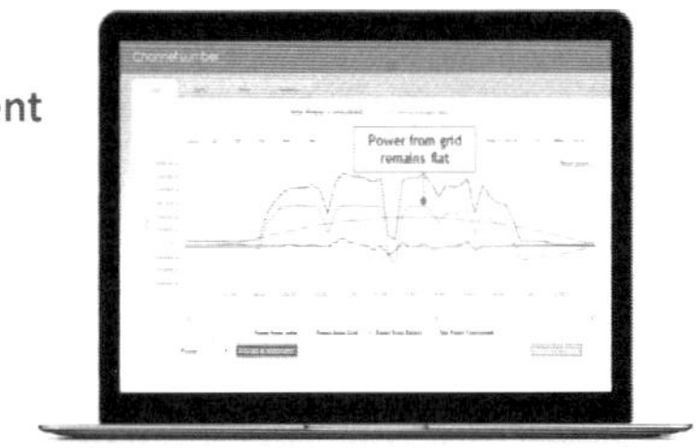

Unprecedented
Safety And Security

Intelligent Control, Modularity
& Expansion Capabilities

Platform Intelligence with
Smart Dispatch Capabilities

NantEnergy

NantEnergy

Air in Resiliency & Demand Management Applications

Energy Storage with Solar roof
Benefit stacking - Demand Management and Resiliency
Central America

NantEnergy — Removing Powerlines From Great Smoky Mountains Nat'l Park

Challenge

- Duke Energy is responsible for servicing the Mt. Sterling Fire Lookout Tower located in the Great Smoky Mountains National Park. The 60-foot tower is a vital component of the park's emergency communication system and must meet rigorous reliability standards.
- A lone 3.5-mile-long powerline connected the tower to the electric grid of Haywood County, N.C. and it fell to Duke to keep vegetation away from the powerlines.
- To do maintenance in the steep terrain of Mt. Sterling backcountry, supplies would often need to be flown in by helicopter or hauled up by pack mule.

Solution

- Duke hired NantEnergy and Sky Renewable Energy & Industrial Solar Consulting to design and build a solar+storage microgrid.
- This gave the communications tower the power it needed to keep the radio on and also granted it energy independence.
- The communications tower is now powered by a 10kW solar array paired with a NantEnergy 95kWh Zinc-air battery that was installed in Spring 2017.

NantEnergy — Removing Powerlines From Great Smoky Mountains Nat'l Park

Results

- The powerlines were removed in November 2018, and the land is now left to return to a natural state.
- Duke saves money on reduced maintenance and the National Park Service has improved power reliability at an important communications tower.

> Having the line going up the mountain for that long, through all that vegetation in the national park, there were a lot of reliability issues, and these issues were triggered when there was an outage
>
> ***Sherif Abdelrazek***
> *Senior engineer at Duke Energy, the utility servicing the line*

Vegetation surrounds the microgrid installed on Mt. Sterling to power its radio relay communications tower

- Customer Category: *Telecom*
- Partner: *Duke Energy & Sky Renewable Energy & Industrial Solar Consulting*
- System Size: *10kW solar array, 95kWh Zn-Air battery*
- Battery Type: *Zn-Air Hybrid Storage*

NantEnergy

Improving Quality of Life in Rural Madagascar

Challenge

- Madagascar, the globe's fourth largest island, is home to nearly 26 million people - only 23% of whom have access to electricity.
- In rural areas, only about 5% of residents have electricity.

Solution

- In 2017, NantEnergy partnered with local Caterpillar dealer Henri Fraise Fils & Cie to develop a solar+storage microgrid in Belobaka, a village 234 miles (376 km) northwest of Madagascar's capital Antananarivo.
- In the first phase of the $1.5 million project, the microgrid is designed to generate and deliver electricity for the first time to 400 households and local businesses.
- The microgrid includes 83kW (peak) solar panels, 16 Zn-Air battery storage systems totaling 315 kWh (48V), and 100 power poles to carry 400 meters of low-voltage lines.

NantEnergy

Improving Quality of Life in Rural Madagascar

Result

- This is an excellent example of a commercial, pay-as-you-go business model in a developing country.
- Using a cell phone SIM card or rechargeable card customers prepay US$0.43 per kWh.
- In addition to providing electric power for threshing machines, lights and other uses, the project also generated several dozen local jobs.

"Having served people in these rural, farming communities for nearly 100 years, we knew how much they needed electrification, and that they would pay for it if we could deliver it affordably. Without NantEnergy's zinc-air batteries, the project economics wouldn't have worked; they are ideal for the climate and longevity as well as cost."

Nicolas Verbert
Vice President of Henri Fraise.

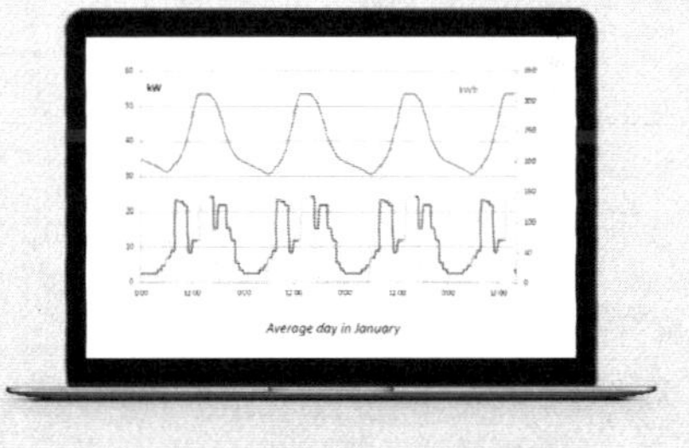

Microgrid Design: Simulations in High Demand

- Customer Category: Remote microgrid
- Partner: Henri Fraise Fils & Cie
- System Size: 83kW (peak) solar panels, 16 battery storage systems totaling 315kWh (48V)
- Battery Type: Zn-Air Hybrid Storage

NantEnergy

Zinc Batteries - Summary

- **Cost** - Metal-Air, and in particular, Zinc-Air technology has a tremendous cost advantage due to low cost of raw materials.
- **Safe and free from toxic materials** – Safest chemistry with no chance of thermal runaway, water based chemistry, No heavy metals like Lead or Cadmium
- **Green** - Storage with one of the least impact to the environment
- **Sustainable** - No resource limitations - Carbon, Zinc, Air are earth abundant and commodity materials
- **Scaleable & Global** - Both technology and manufacturing are scaleable and can be made locally anywhere with low manufacturing cost
- **Robust** - Wide range of temperature, No limit to depth of discharge
- **Long Duration** - Can enable cost-effective long duration storage, eliminating the need for diesel generators to provide grid reliability and resilience
- **Proven** – Technology already deployed and proven for 6+ years

Zinpro Corporation

The Zinpro Story

"There's Something About Zinc"

1965

Dean Anderson

"There is something about Zinc!!"

ZINPRO
PERFORMANCE MINERALS®

1976

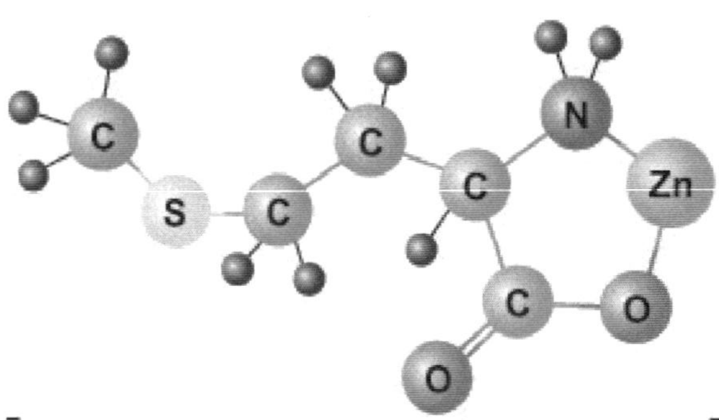

Zinpro's History

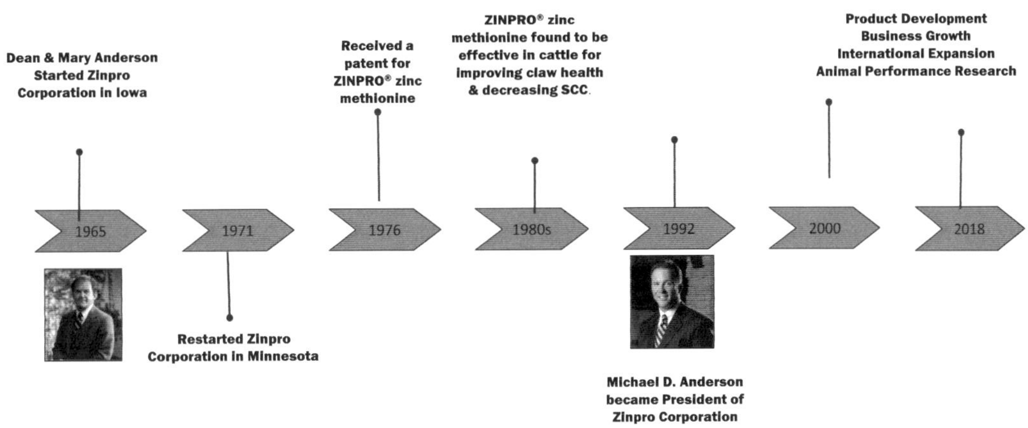

The Last 50 Years

What Have We learned?

Animal Health & Performance

- Enhanced Immune Function
- Disease Mitigation
- Hoof & Hair Coat
- Reproductive Efficiency
- Epithelial Integrity
- Reduction in Stress
- Enhanced Health & Performance
- Improvement in Feed Efficiency ~3%

2018 Global Feed Tonnage

2018 - 1.103 billion metric tons of feed

-3%

That's Enough Feed For 6.9 Billion Broilers Per Year

THAT'S 1 CHICKEN FOR EVERY PERSON ALIVE

67.8 Million lbs.
Additional Beef =
485k Additional Animals

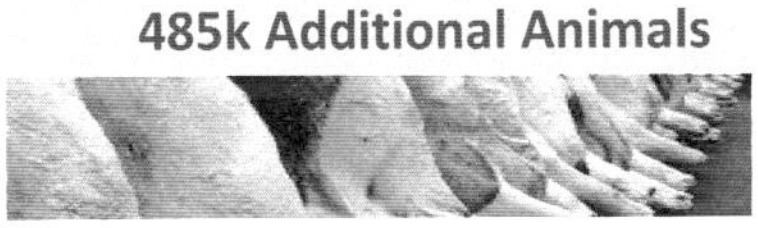

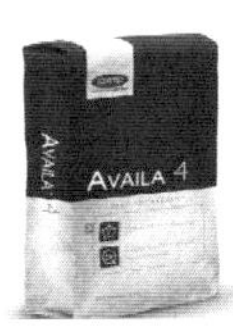

Potential Savings of 1.73 bu.
Corn/Head,
32.2 Million bu/yr
195,000 acres/yr

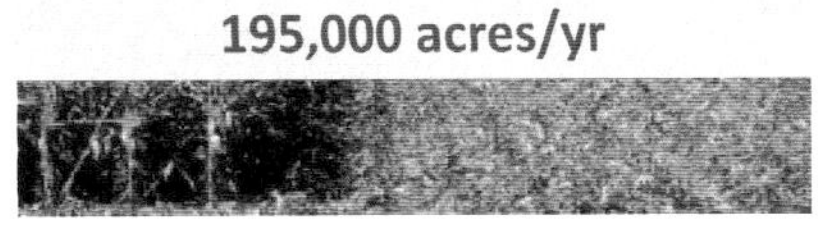

Potential Savings of 100
Trillion Gallons of
H20/Year

Potential Savings of
38.6 Million lbs. of
Nitrogen/Year

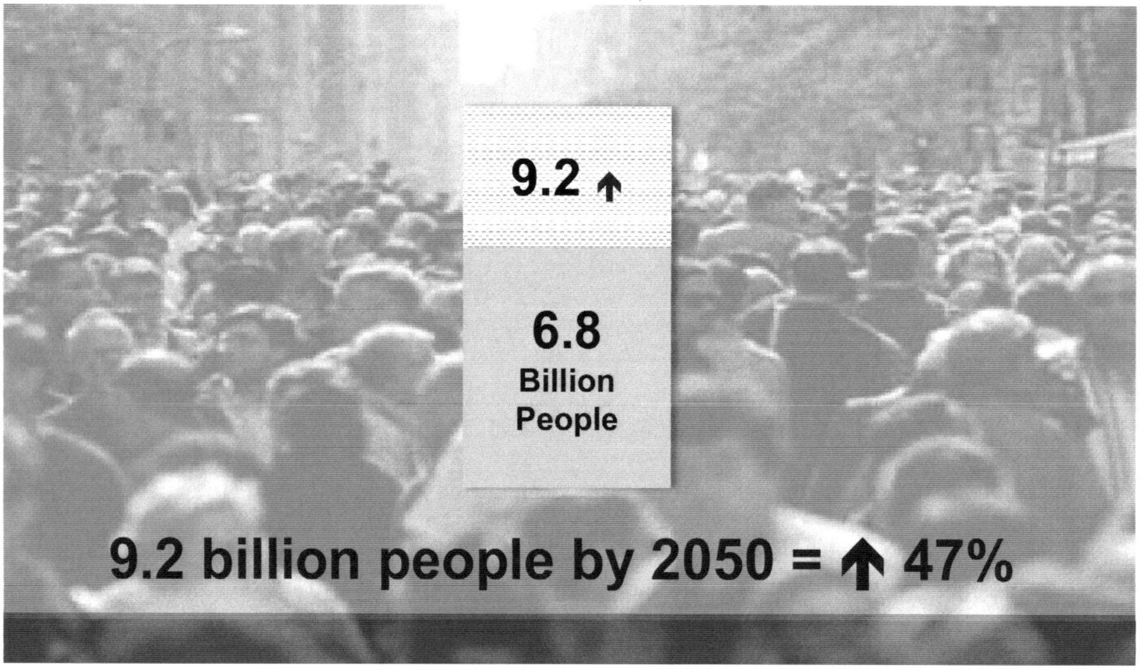
9.2 ↑
6.8
Billion
People
9.2 billion people by 2050 = ↑ 47%

161,000 New People Every Day

The Last 50 Years

What Have We learned?

Then

"The essentiality of zinc in humans was established in 1963"

"In 1963, we knew of only 3 enzymes that required zinc for their activities"

Now

For humans, it is estimated that about 3000 proteins contain zinc as a cofactor.

"Today >300 enzymes and >1000 transcription factors are known to require zinc for their activities"

Ananda S. Prasad
Wayne State University School of Medicine, Department of Oncology, and Karmanos Cancer Center, Detroit, MI, 48201

Zinc has critical roles in every cell, both as a structural element and a dynamic signaling ion

ZINPRO
PERFORMANCE MINERALS®

http://iszb.org/why-zinc

“Zinc is like <u>no other</u> nutrient”

IN THE WAY IT IS REGULATED, STORED, UTILIZED AND NEEDED BY THE BODY

Janet C. King, Ph.D., R.D.
Research Professor Emerita, Department of Nutrition
Senior Scientist, Children's Hospital Oakland Research Institute

Given the fundamental functions of zinc it should come as no surprise that *Zinc* is implicated in:

- **Alzheimer disease**
- **Stroke**
- **Diabetes**
- **Cancer**
- **Parkinsonism**
- **Autism Spectrum Disorder**
- **More**

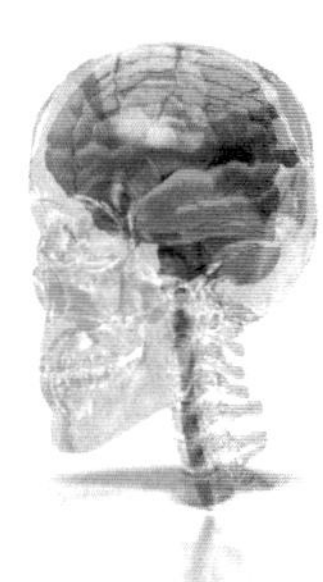

CDC increases estimate of autism's prevalence by 15 percent, to 1 in 59 children

Autism Speaks calls on nation's leaders to adequately fund critically needed research and support services

April 26, 2018

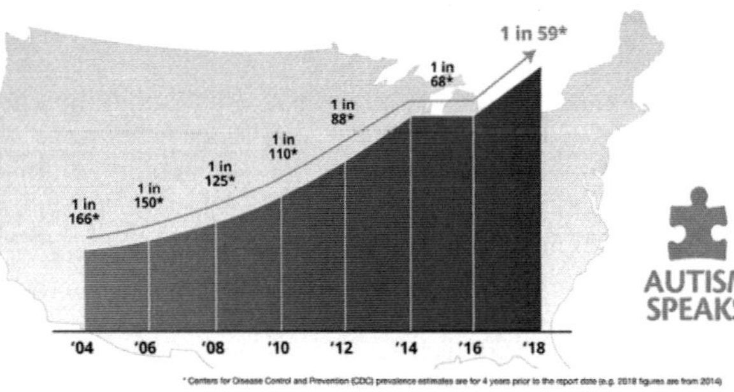

AUTISM SPEAKS®

Arora M, Reichenberg A, Willfors C, Austin C, Gennings C, Berggren S, Lichtenstein P, Anckarsäter H, Tammimies K, Bölte S. Fetal and postnatal metal dysregulation in autism. Nat Commun 2017; 8: 15493.

Skalny AV, Simashkova NV, Klyushnik TP, Grabeklis AR, Bjørklund G, Skalnaya MG, Nikonorov AA, Tinkov AA. Hair toxic and essential trace elements in children with autism spectrum disorder. Metab Brain Dis 2017; 32 (1): 195-202.

Babaknejad N, Sayehmiri F, Sayehmiri K, Mohamadkhani A, Bahrami S. The Relationship between Zinc Levels and Autism: A Systematic Review and Meta-analysis. Iran J Child Neurol 2016; 10 (4): 1-9.

Crăciun EC, Bjørklund G, Tinkov AA, Urbina MA, Skalny AV, Rad F, Dronca E. Evaluation of whole blood zinc and copper levels in children with autism spectrum disorder. Metab Brain Dis 2016; 31 (4): 887-90.

Tabatadze T, Zhorzholiani L, Kherkheulidze M, Kandelaki E, Ivanashvili T … AUTISM SPECTRUM DISORDER. Georgian Med News 2015; (248): 7…

Mac…

Li SC…

Yasu…

Al-F…

Russ… …, Jones BM, Languell PA, Lyman PJ, March KP, Meuer … Warren JS. Plasma copper and zinc concentration in …

Blau… … of Symptoms among Children with Autism. Maedica …

Russo AJ, Devito R. Analysis of Copper and Zinc Plasma Concentration and the Efficacy of Zinc Therapy in Individuals with Asperger's Syndrome, Pervasive Developmental Disorder Not Otherwise Specified (PDD-NOS) and Autism. Biomark Insights 2011; 6: 127-33.

Lakshmi Priya MD, Geetha A. Level of trace elements (copper, zinc, magnesium and selenium) and toxic elements (lead and mercury) in the hair and nail of children with autism. Biol Trace Elem Res 2011; 142 (2): 148-58.

Blaurock-Busch E, Amin OR, Rabah T. Heavy metals and trace elements in hair and urine of a sample of arab children with autistic spectrum disorder. Maedica (Buchar) 2011; 6 (4): 247-57.

Yasuda H, Yoshida K, Yasuda Y, Tsutsui T. Infantile zinc deficiency: association with autism spectrum disorders. Sci Rep 2011; 1: 129.

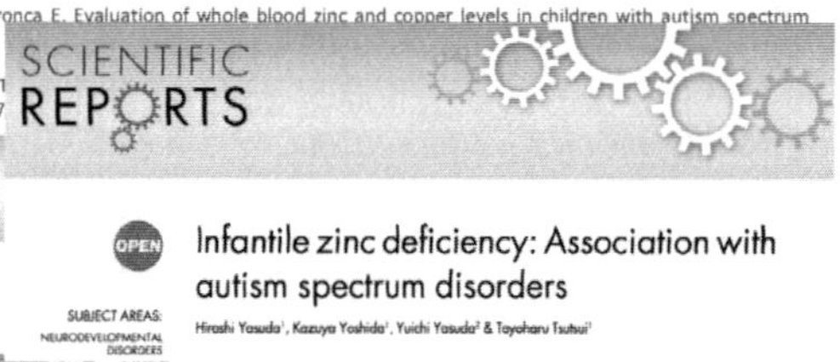

SCIENTIFIC REPORTS

SUBJECT AREAS: NEURODEVELOPMENTAL DISORDERS

OPEN

Infantile zinc deficiency: Association with autism spectrum disorders

Hiroshi Yasuda[1], Kazuya Yoshida[1], Yuichi Yasuda[2] & Toyoharu Tsutsui[1]

nature COMMUNICATIONS

ARTICLE

Received 28 Apr 2016 | Accepted 4 Apr 2017 | Published 1 Jun 2017

Fetal and postnatal metal dysregulation in autism

Manish Arora[1], Abraham Reichenberg[1,2], Charlotte Willfors[3,4], Christine Austin[1], Chris Gennings[1], Steve Berggren[3,4], Paul Lichtenstein[5], Henrik Anckarsäter[6], Kristiina Tammimies[3,4] & Sven Bölte[3,4]

"Zinc in Gut-Brain Interaction Neurological Disord

Hindawi Publishing Corporation
Neural Plasticity
Volume 2015, Article ID 972791,
http://dx.doi.org/10.1155/2015/97

Review Article
Zinc in Gut-B
Neurological

frontiers
in Behavioral Neuroscience

http://m.info7.mx/a/noticia/658119

Gender Dependent Evaluation of Autism like Behavior in Mice Exposed to Prenatal Zinc Deficiency

*Stefanie Grabrucker[1], Tobias M. Boeckers[1] and Andreas M. Grabrucker[1,2]**

[1] Institute for Anatomy and Cell Biology, Ulm University, Ulm, Germany, [2] WG Molecular Analysis of Synaptopathies, Neurology Department, Neurocenter of Ulm University, Ulm, Germany

Zinc deficiency has recently been linked to the etiology of autism spectrum disorders (ASD) as environmental risk factor. With an estimated 17% of the world population being at risk of zinc deficiency, especially zinc deficiency during pregnancy might be a common occurrence, also in industrialized nations. On molecular level, zinc deficiency has been

"Zinc deficiency is responsible for 40% of all nutritionally related deaths in children under the age of five"

Dr. Janet King, Children's Hospital Oakland Research

~500,000 Children Per Year

THE NEXT 50 YEARS

"It is paradoxical, yet true, to say, that the more we know, the more ignorant we become in the absolute sense..............."

Nikola Tesla – 1856-1943

"There's Something About Zinc"

ZINPRO

PERFORMANCE MINERALS®

A tale of two markets: Concentrates abound while metal market remains tight

Ryan Cochrane
CRU Group

ryan.cochrane@crugroup.com

AZA – 2019

February 2019

Zinc market has a history of price 'spikes' where miners have progressively carved out a larger share of zinc revenues

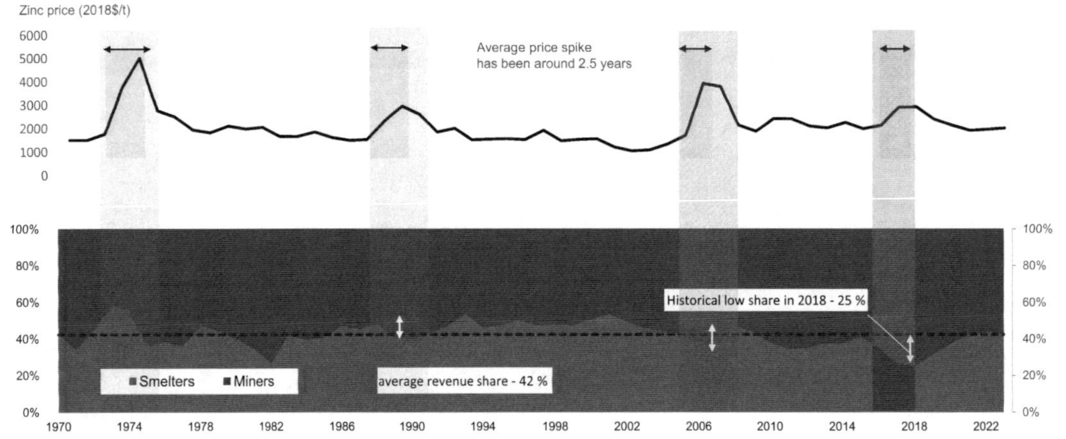

Data: CRU, LME

2

What about past supply – greatest supply volatility seen for Ni, least for Cu. Supply shocks (negative and positive) common for Nickel and Pb/Zn

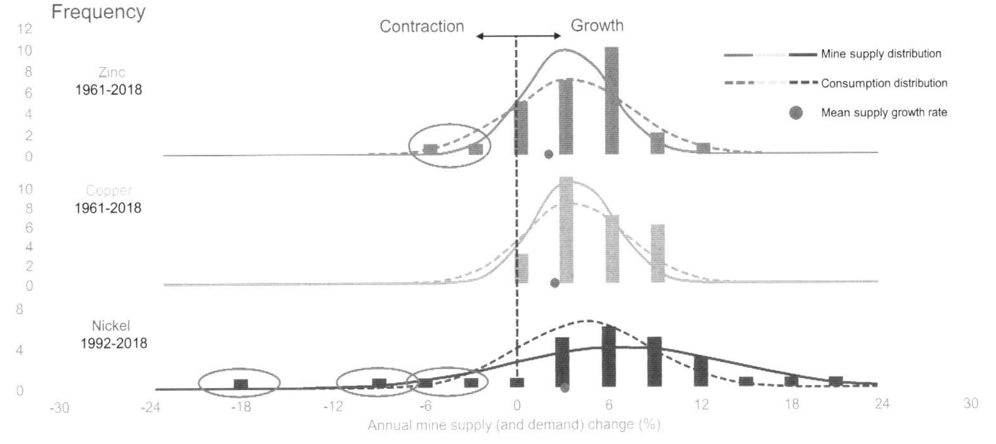

Data: CRU

3

Most recent cycle has demonstrated strong supply elasticity in zinc market

Lead/Zinc –
Polymetallic sector with relatively small number of substantial mines, and a high proportion of small mines.

- Naturally high depletion (burn) rates
- Price related project restarts (and closures) high in China and outside of China.
- Project capital requirements relatively modest
- Holders of limited number of major assets can exert significant 'strategic' effect on market

Data: CRU

4

Agenda

Mine supply and project overview

Refined supply and demand outlook

Balances, prices and indicators

Mine supply to see strongest gains this year and 2020 before attrition begins to offset new gains

Mine supply forecast vs. potential

million tonnes zinc

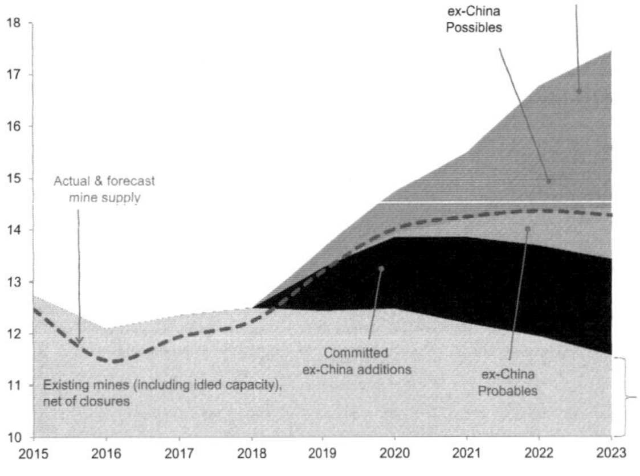

Committed mine supply - >800,000 t/y at risk

2018-23 production loss; '000 tonnes zinc

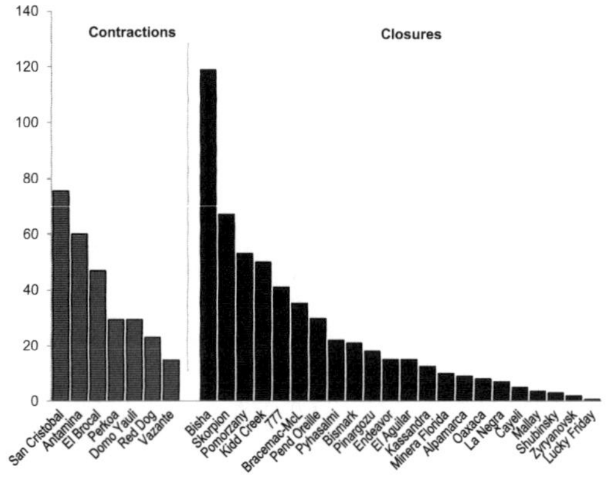

Data: CRU

Largest wedge of mine supply is expected from committed plans outside of China

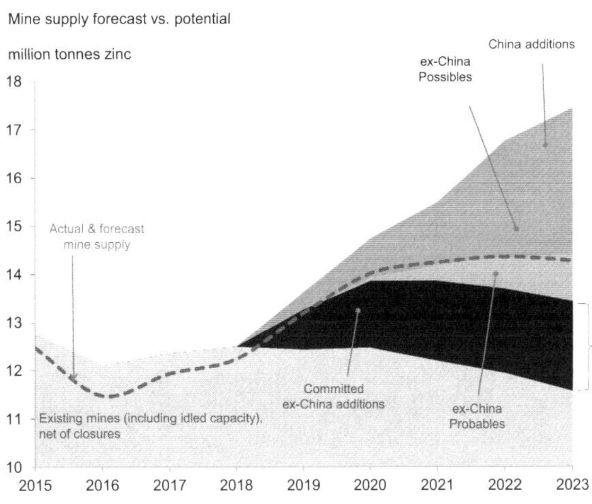

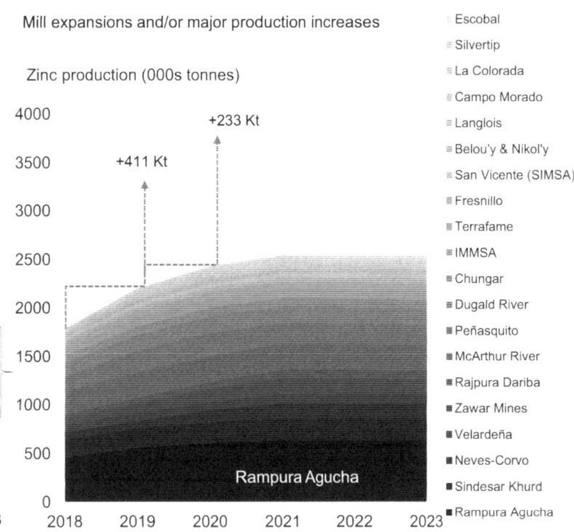

Data: CRU

7

Largest wedge of mine supply is expected from committed plans outside of China

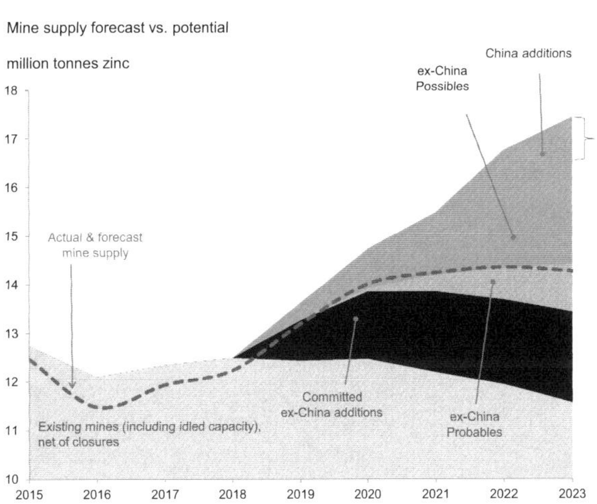

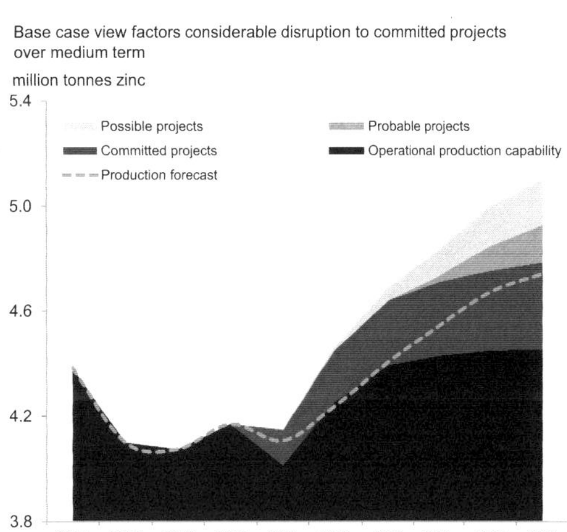

Data: CRU

8

Largest wedge of mine supply is expected from committed plans outside of China

Mine supply forecast vs. potential

million tonnes zinc

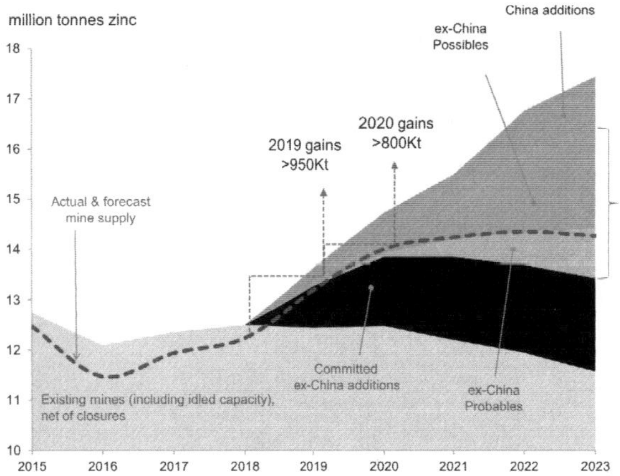

Uncommitted mine supply

potential 2023 production; '000 tonnes zinc

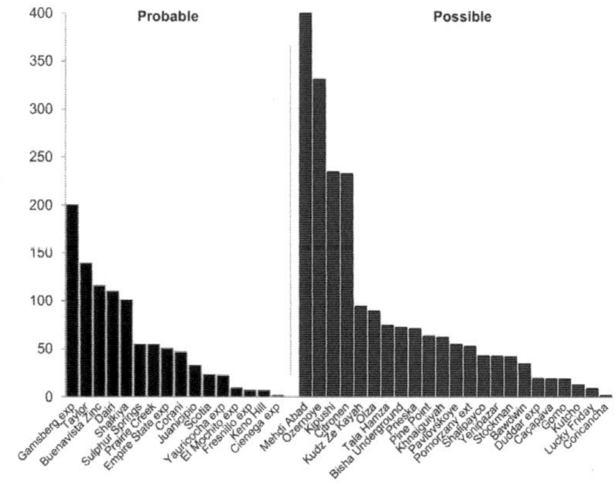

Data: CRU

Zinc projects remain the least capital intensive option compared to Copper and Nickel

Zinc projects remain least capital intensive – after adjustment for metal price

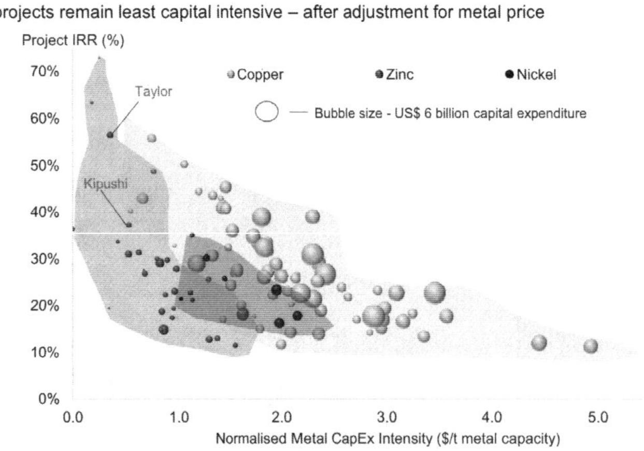

Data: CRU

Risks to our supply outlook

We apportion just 10-15% of supply from possible and probable projects to our balance.

- This results in just over 300,000 tonnes of supply being derived from our population of possible and probable projects.
- In the event that Mehdiabad (which we list as possible but understand is under construction) supplies at full capacity (400,000 t/y), this alone would more than outweigh our total allowance for uncommitted projects.
- Big-ticket mining projects include Gamsberg Phase II, Mehdiabad, Taylor, Ozernoye, Dairi, Shalkiya, Kipushi, Buenavista and Huoshaoyun.

Agenda

Mine supply and project overview

Refined supply and demand outlook

Balances, prices and indicators

11

Smelting capacity is not the short term issue, utilisation rates – particularly in China will be key

Global demand growth will continue to trend lower

refined metal demand, million tonnes

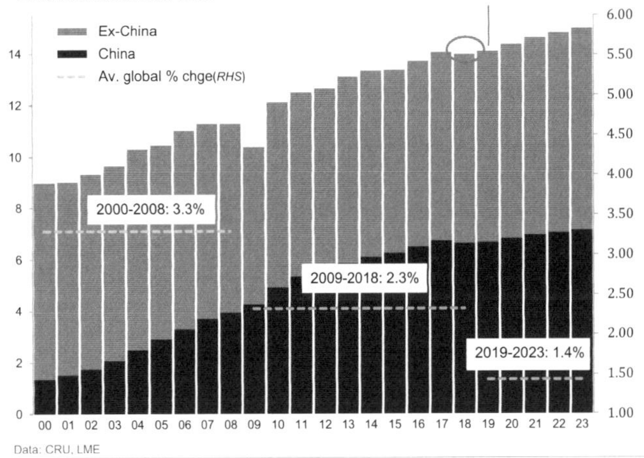

Data: CRU, LME

Weak demand in 2018, but set to improve modestly in 2019

Chinese demand growth disappointed in 2018 and we now have it falling by 1.5% year-on-year:

- Galvanised sheet output was hit by trade tariffs and the slowdown in China's automotive sector. Our Steel Team has China's galvanised sheet output down by 1% y-o-y.
- Die casters in Guangdong province reported steep declines in demand in both domestic & export markets
- End-use demand continues to be disrupted by environmental issues.
- Chinese demand should return to growth this year as stimulus begins to take effect.

- In the world ex-China we have made downward adjustments to our expectation of demand growth in line with macro economic and political headwinds

12

Smelting capacity is not the short term issue; utilisation rates, particularly in China, will be key

Smelter production forecast vs. capacity
million t/y

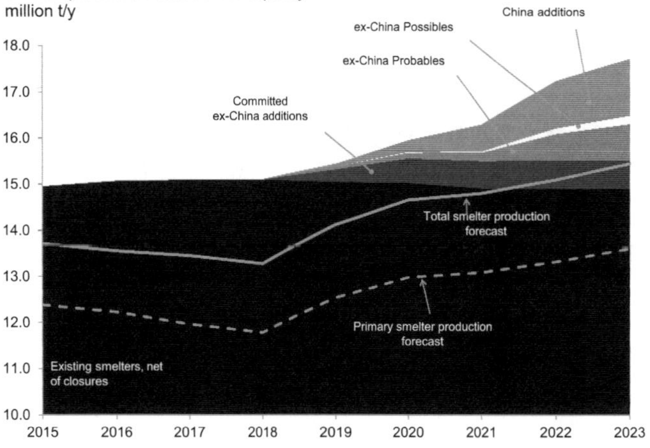

Data: CRU, LME

Smelter capacity and utilisation

China smelting capacity - we have pushed back a number of Chinese smelter projects and expect negligible net gains in effective smelting capacity in 2019. Growth is set to return strongly from 2021.

Ex-China smelting capacity changes – Townsville (+50 Kt/y), Torreon (+120 Kt/y), Mooresboro (+150 Kt/y), Chelyabinsk (+15 Kt/y) and HZL (+275Kt/y). UMMC announced Electrozink would be idled (- 95Kt/y),

Utilisation rates will be a key consideration for the metal balance in 2019 – particularly for Chinese smelters

13

Demand growth and minimum revenue levels have been important to high utilisation rates in China

Demand declined in China in every quarter of 2018

year-on-year percentage change in zinc demand

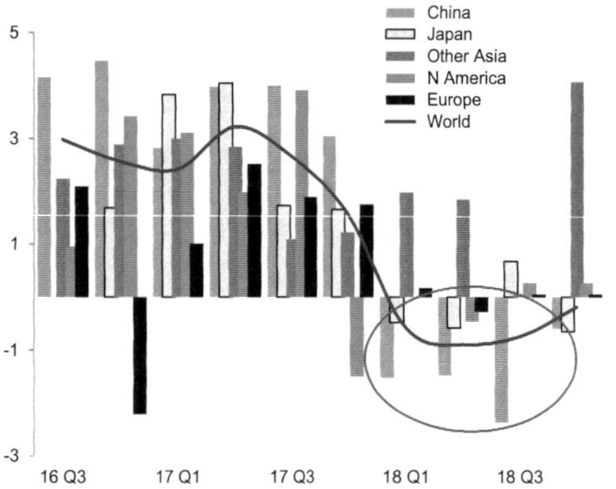

Chinese smelting capacity growth has tracked consumption growth closely although domestic 'bottleneck' has emerged since 2017

000's tonnes zinc

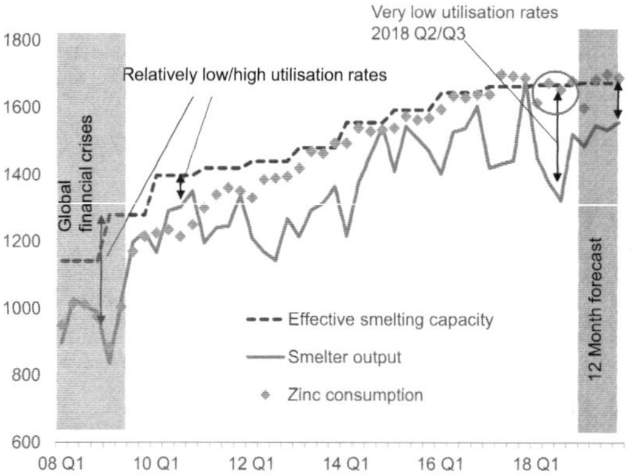

Data: CRU, LME

14

Some major smelters which could raise utilisation rates in 2019

Around 1,040Kt of unused capacity has been identified from 16 major smelters in China.
CRU estimates 392Kt of this may return to market in 2019

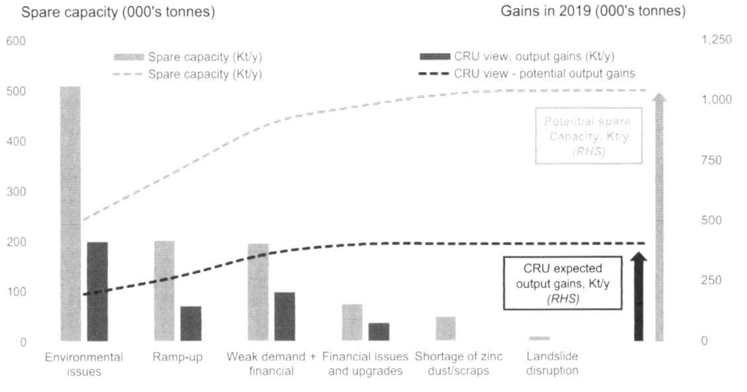

Data: CRU, LME

Historically, the most important indicator has been raw material availability

We expect that gains in concentrate inventories in 2018, alongside significant mine supply growth projected in 2019 and 2020, will drive up utilisation rates
000's tonnes

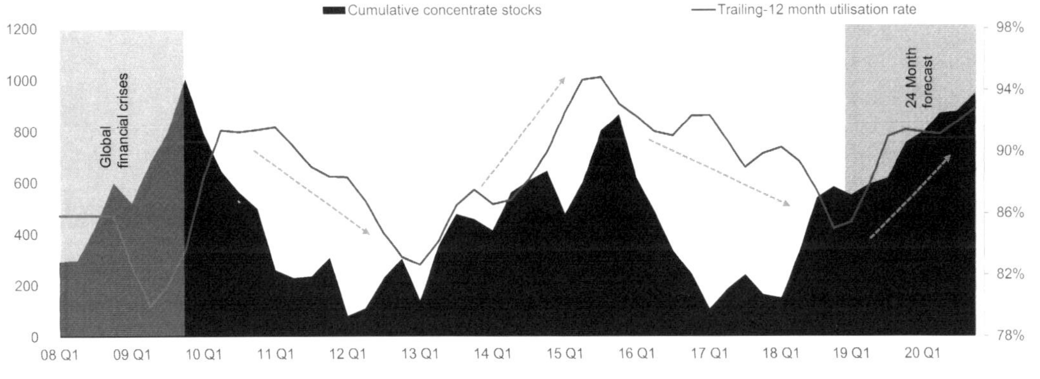

Data: CRU, LME

CRU 50

Agenda

Mine supply and project overview

Refined supply and demand outlook

Balances, prices and indicators

CRU 50

Concentrate market surpluses have begun – rising TCs have marked the turning point.

Concentrate surpluses will grow more strongly than metal surpluses
'000 tonnes contained zinc

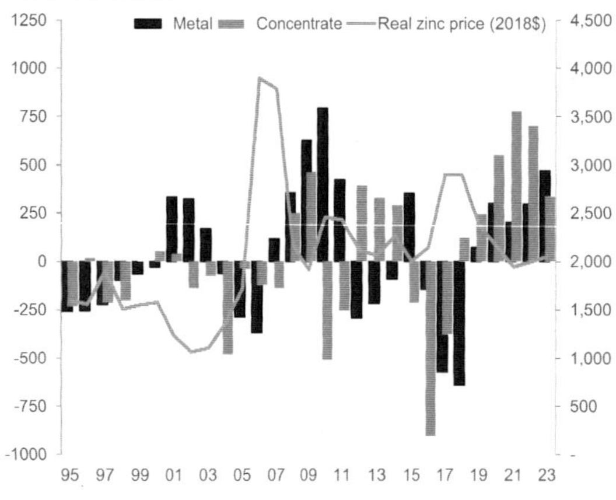

Spot TCs marked the turning point in this cycle

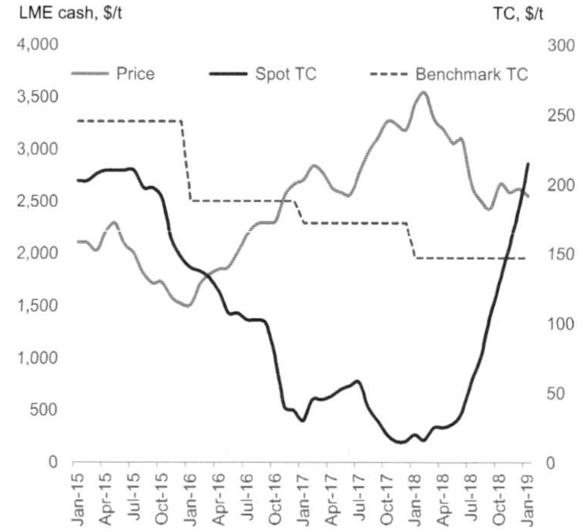

Data: CRU, LME

Concentrate market has eased while metal still tight - we expect a balanced metal market in 2019

Concentrate stock recovery to precede metal stock recovery

cumulative zinc market surplus; adjusted; million tonnes

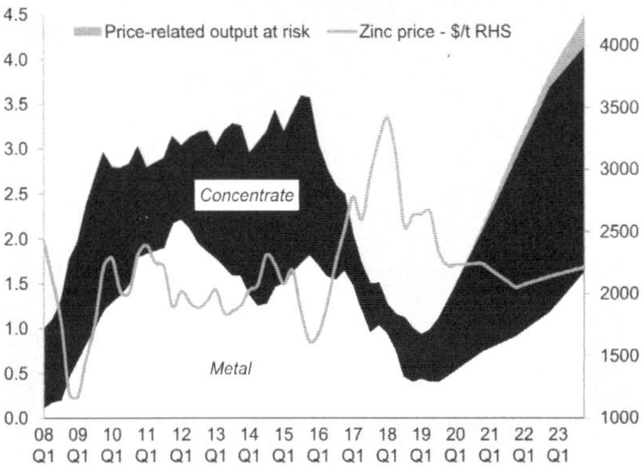

Metal stocks will not reach previous lows and will build quickly from 2020

Metal stock ratios, weeks, LME price $/t

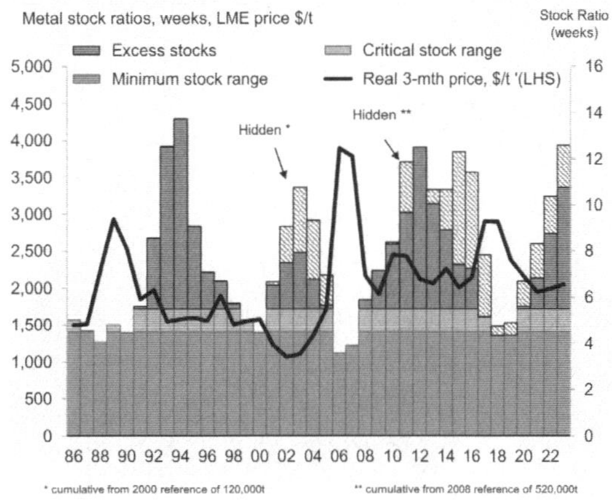

Data: CRU, LME

19

Strong supply variance is expected, with market moving into oversupply before strong project development requirement opens up in mid 2020s

Moving into period of oversupply before looming requirements for new mine supply in mid 2020s

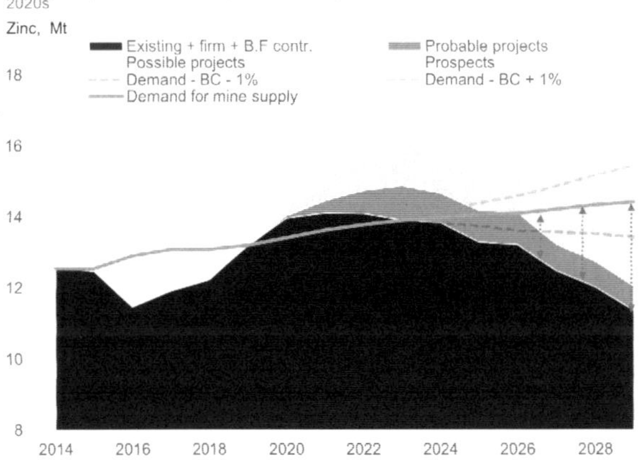

Operating + required projects cash cost projections by 2029

Cash Costs Including By-product Credits $/t

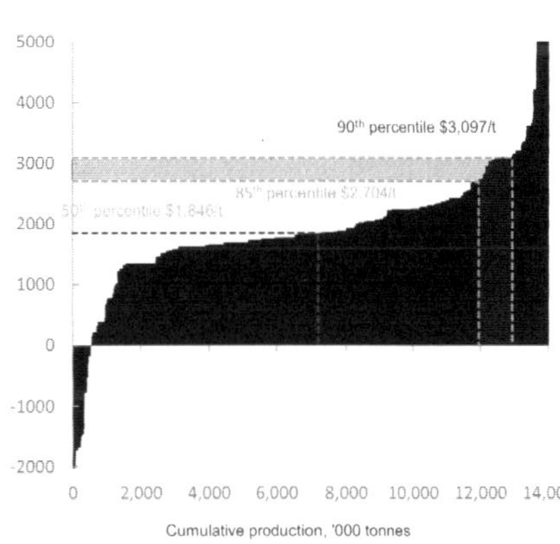

Data: CRU

20

Historically, support found at 50th percentile but prices have tended toward 90th

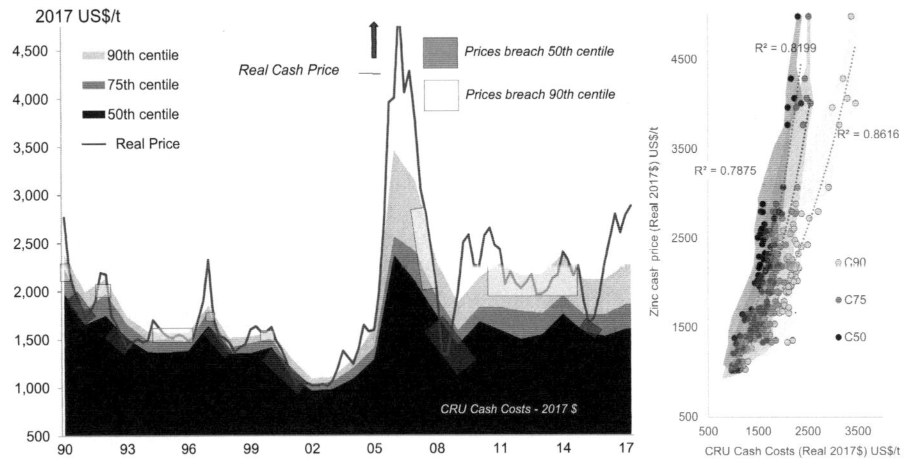

Data: CRU, LME

Summary outlook

Supply

- Concentrate availability has improved as reflected in higher treatment charges
- In the medium term, mine supply will outpace smelter capacity growth, supporting a further increases in TCs and investment in smelter capacity
- There is the potential for a significant build in concentrate stocks over the next 5 years
- By mid-2020s reserve attrition at existing operations will drive a significant requirement for project development

Demand

- Chinese zinc demand, and supply has disappointed in 2018 and both were lower y-o-y
- The expectation of a slowdown in global IP growth suggests that demand will grow only modestly over medium term
- From 2019 onwards, China will continue to add the most to global demand in tonnage terms and we expect global demand to grow to around 15Mt (or 200Kt/y) by 2023

Balance & Price

- Forecasting strong cyclicity in prices from short-medium-longer term
- Short term - tight refined market to support prices through 2019 Q2
- 2019 - smelter capacity at raised utilisation rates will be sufficient to meet modest demand growth, bringing the metal market to balance before surpluses emerge. Prices will need to cut deep into the cost curve to motivate supply cuts to rebalance the market in early 2020s.
- By mid-2020s, reserve attrition is setting the scene for a significant project development requirement which will support LT prices between $2,700-3,000/t

CRU's Lead/Zinc and Precious Metals Team

Dr. Ryan Cochrane
Research Manager: Lead, Zinc and Precious Metals
Ryan.cochrane@crugroup.com
Tel. +44 20 7903 2093

Helen O'Cleary
Zn - Senior Analyst
Helen.ocleary@crugroup.com
Tel. +44 20 7903 2038

Neil Hawkes
Pb - Principal Analyst
Neil.Hawkes@crugroup.com
Tel. +44 20 7903 2101

Dr. Alice Fox
Pb/Zn - Principal Analyst
Alice.fox@crugroup.com
Tel. +44 20 7903 2261

Dr. Kirill Kirilenko
Precious Metals -Senior Analyst
Kirill.Kirilenko@crugroup.com
Tel. +44 20 7903 2282

Christine Meilton
Pb/Zn - Senior Analyst
Christine.meilton@crugroup.com
Tel. +44 20 7903 2098

Brian Peng
Pb/Zn - Research Analyst - **Beijing**
Brian.peng@crugroup.com
Tel. 86-10-65102206-235

Dina Yu
Pb/Zn - Senior Analyst - **Beijing**
Dina.yu@crugroup.com
Tel. 86-10-65102206-352

Dimitris Zomas
Pb/Zn - Research Analyst
Dimitris.zomas@crugroup.com
Tel. +44 20 7903 2232

CRU International Limited
London | Sydney | Tokyo | Beijing | Shanghai | Hong Kong | Mumbai | New York | Pittsburgh | Sao Paulo | Santiago

Registered in England No.940750. Registered office: Chancery House, 53-64 Chancery Lane, London, WC2A 1QS, UK

Legal notice

This presentation is private and confidential. It must not be disclosed in whole or in part, directly or indirectly or in any other format without the prior written permission of CRU International Limited.

CRU International Limited's responsibility is solely to its clients and its liability is limited to the amount of the fees actually paid for professional services.

Although reasonable care and diligence has been used in the preparation of this presentation, we do not guarantee the accuracy of any data, assumptions, forecasts or other forward-looking statements. We accept no liability to third parties, howsoever arising.

CRU takes information security seriously and currently holds the UK Government approved Cyber Essentials certification. This certifies that we have the appropriate security controls across our organisation and third party suppliers to protect our information assets. CRU also has a privacy policy in place which explains how we handle personal data on our customers.

About CRU

CRU's reputation with customers across **mining**, **metals** and **fertilizers** is for integrity, reliability, independence and authority.

CRU's insights are built on a twin commitment to quality primary research and robust, transparent methodologies.

CRU invests in a global team of analysts, the key to gaining a real understanding of critical hard-to-reach markets such as China.

We strive to provide customers with the best service and closest contact – flexible, personal, responsive.

CRU – big enough to deliver, a high quality service, small enough to care about all our customers.

CRU's office locations

25

PURSELL
AGRI-TECH
SINCE 1904

The Importance of Zinc in Food Security and Human Health

February 27, 2019

PURSELL
AGRI-TECH
SINCE 1904

Our Mission

Develop and commercialize next generation fertilizers to increase yields, enhance human nutrition, and improve environmental impact to water and climate.

PURSELL AGRI-TECH CONFIDENTIAL

Zinc Nutrient Initiative – Huge Success

PURSELL AGRI-TECH SINCE 1904

Zinc is, by far, the best known of all micronutrients due to the excellent job the IZA is doing with its Zinc Nutrient Initiative.

The agricultural industry is well aware of the benefits of applying zinc to improving crop yields and human health.

Because of the professionalism of your efforts, IZA has a seat at the table at the United Nations, Gates Foundation, USAID and other critical organizations focused on human health.

Zinc's Importance in Agriculture:

Food Security (hunger)

Yield increases in soils deficient in Zn

Malnutrition (health)

Improving public health and child morbidity

Annual sales of fertilizer are greater than sales of seeds, ag chemicals, and machinery spending combined

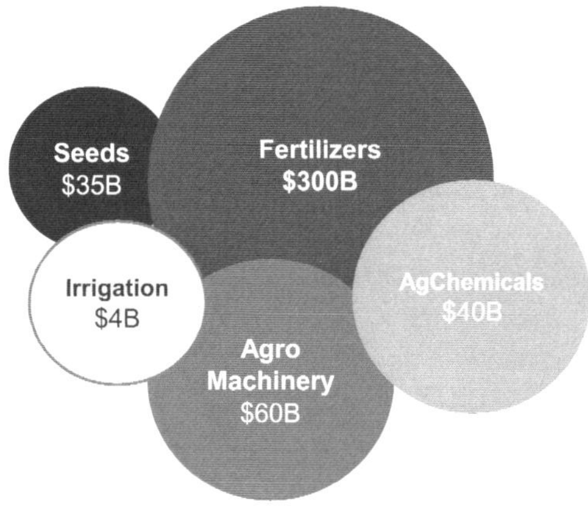

PURSELL AGRI-TECH CONFIDENTIAL

2018 United Nations Sustainable Development Goals – Importance of Zinc

#1 End poverty in all forms everywhere

#2 End hunger, achieve food security, and improved nutrition and promote sustainable agriculture

- By 2030 end all forms of malnutrition
- By 2025 achieve targets on reducing stunting and wasting in children under 5 years of age

There are 7.6 billion
people on Earth.

Countdown to 2050

PURSELL
AGRI-TECH
SINCE 1904

There will be **9.6 billion people** in 2050, eating as much food as **13 billion people** at today's intake levels

2 billion more people
2X as much food
1X arable land

PURSELL AGRI-TECH CONFIDENTIAL

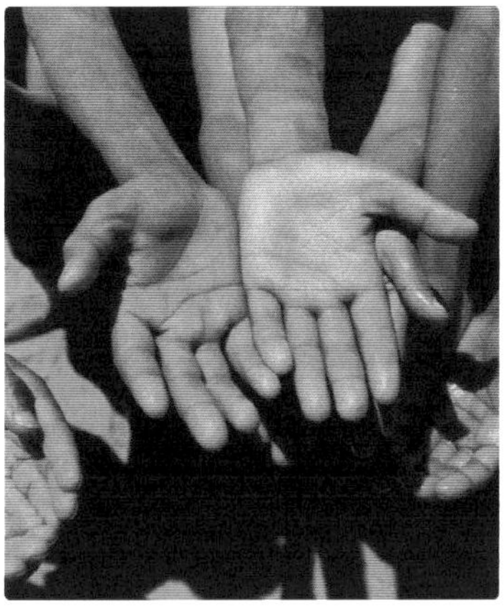

1.02 billion people
suffer from hunger.

That's more people than live in the United States, Canada and the European Union combined.

Source: State of Food Insecurity in the World, 2008 FAO."Food Security Statistics"

Global Hunger Distribution

Hunger is found most in the developing world, where 95% of births take place.

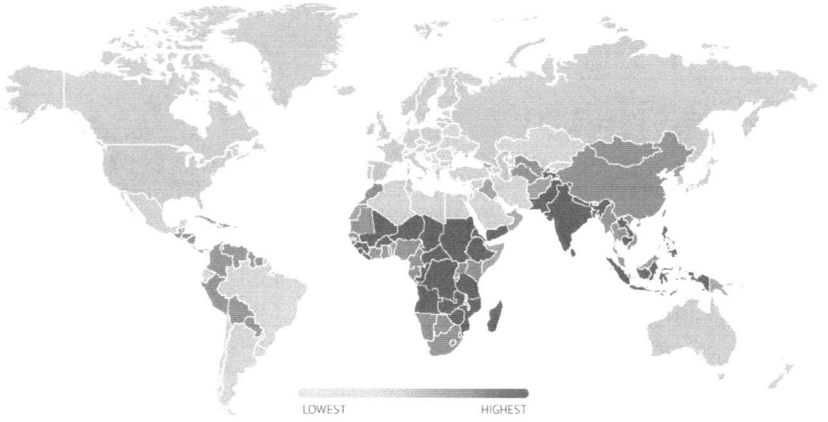

Source: International Food Policy Research Institute

"Hunger" now includes "hunger and nutrition".

Focus changing from just calories to a nutritious diet.

Hidden Hunger:

Billions of people worldwide who suffer from lack of **essential nutrients** like zinc, vitamin A, and iron.

Malnutrition is a significant global driver of death, disease, and stunted growth & cognitive development

25,000 people will die today from malnutrition.

More than half will be young children.

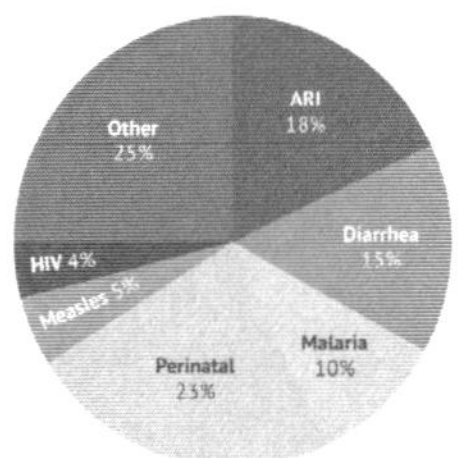

Deaths Associated With Hunger

Source: World Health Organization

Nine-year-old children in Guatemala are significantly shorter than the World Health Organization's global average height for their age. Chronic malnourishment, which causes stunted growth, is an epidemic in Guatemala. In rural villages, more than 80 percent of the population is stunted. (ABCNews)

Nutrient Malnutrition:

- Lowered IQ
- Stunted Growth
- Blindness
- Increased disease susceptibility
- Increased risk for mothers
- Death

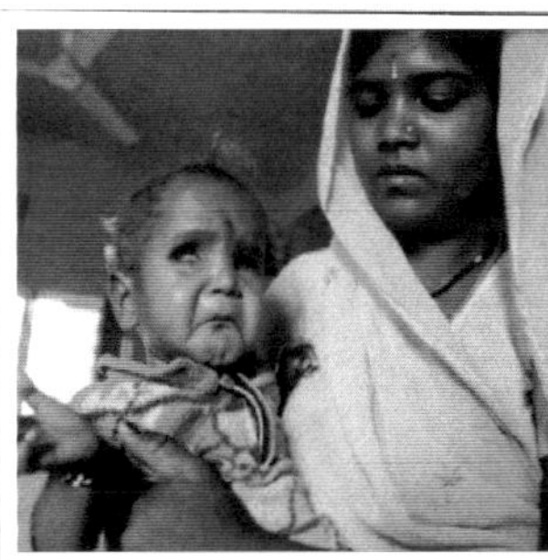

Sheela Adivasi holds her 18-month-old son Deepak at a Nutrition Rehabilitation Centre in Badarwas block, in the central Indian state of Madhya Pradesh, March 7, 2008. A couple of months ago, Adivasi's malnourished infant son fell sick and his eyes filled with pus. By the time the infection cleared up, Deepak's pupils had turned a pearly white. He is now permanently blind. **Had he not been so deprived of vitamins, he probably would have suffered only an itchy but harmless bout of pink eye.** REUTERS/Raj Patidar (INDIA)

Source: http://www.trust.org/alertnet/news/factbox-whats-the-impact-of-hidden-hunger

Hidden Hunger affects the developing world.

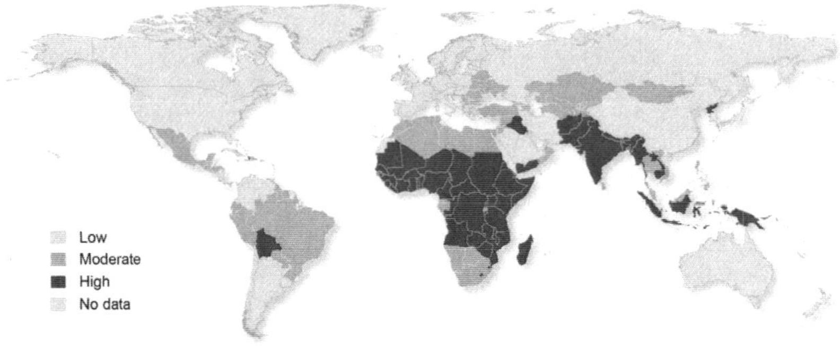

Source: HarvestPlus.org

Locations of Zinc deficiency

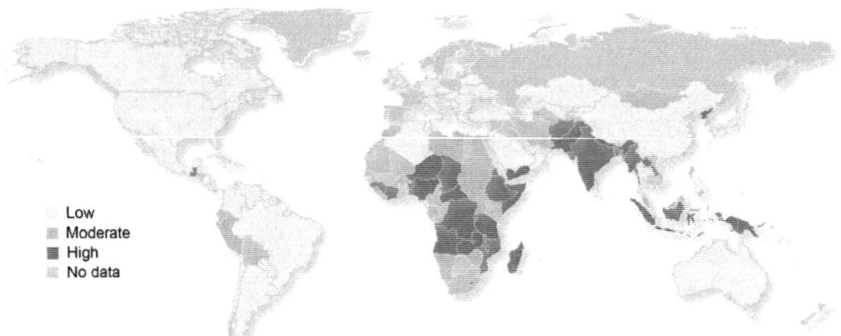

Source: HarvestPlus.org

Zinc is involved in more bodily functions than any other mineral

- Essential to 300 enzyme systems, normal growth, sexual function, vision, immune system
- Intellectual capability

Zinc malnutrition effects

- Stunted growth and development
- Skin lesions
- Diarrhea
- Organ malfunction
- 450,000 children die every year due to zinc deficiency
- 800,000 people at risk of dying each year due to zinc deficiency

Zinc is an also an essential micronutrient for plant growth

- Deficiencies in zinc can result in crops not achieving maximum yield potential

Nutrients added to the soil can be absorbed by the plants and enter the food.

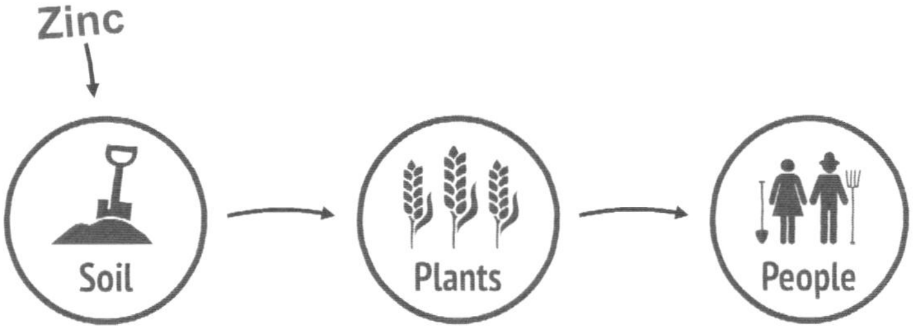

Benefits of our Seed Core Technology

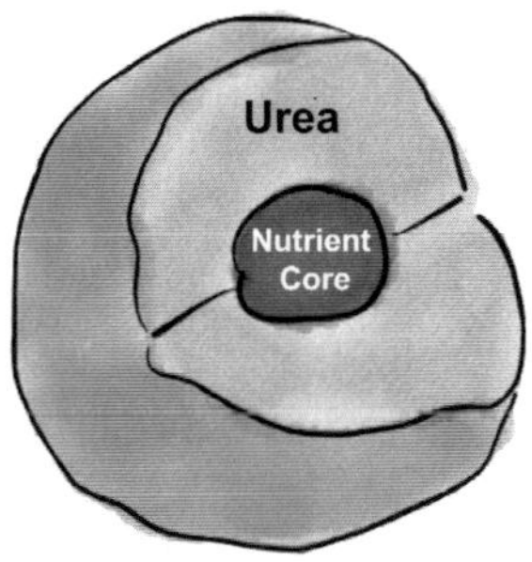

- Uses established distribution and utilization/application channels.
- Improves human health
- Improves yields of plants where soils are deficient
- Improves the environment by being more efficient

Fluid Bed Granulation

We swap the urea cores with our customizable micronutrient cores.

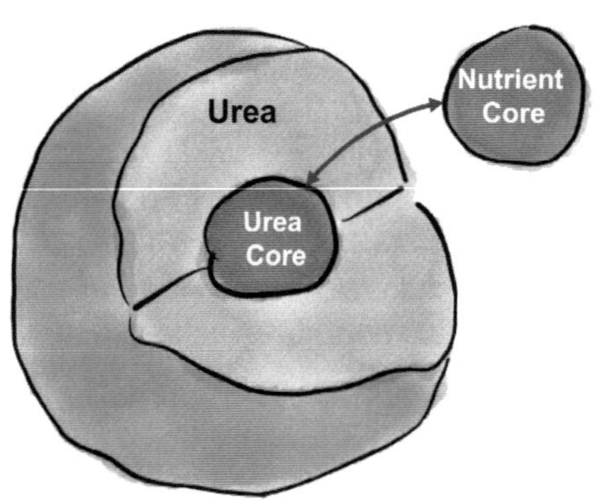

- $ZnSO_4$ or ZnO
- Fe
- Vitamin A
- Iodine

Bangladesh Agricultural Research Institute Zinc-Crops Project on Yield and Zinc Uptake by Maize

Treatment	Grain Yield (t/ha)	% yield increase	Zn content (ppm)	% Zn content increase
Zn- urea briquette	7.71	6.30	25.34	49.1
Zn- urea prilled	7.51	3.60	25.22	48.4
Zn core urea 0.5%	8.48	17.0	25.74	51.4
Zn core urea 1%	8.74	20.6	25.80	51.8
Control	7.25	-	17.00	-
SE(±)	0.24	-	1.9	-

Variety: : **BARI Hybrid Maize 9,** Location: **Domar, Nilphamari(AEZ-3)**

Civilization as it is known today could not have evolved, nor can it survive, without an adequate food supply.

- Norman Borlaug

The Uber of **EV-**erything

TED GRAHAM
Head of Open Innovation
General Motors
@tedgraham

September 1996 $2.95
True North
VOLLEYBALL
Magazine
1996 Olympic
Bronze Medalists
Mark Heese and
John Child

FBI

A24

Brian Chesky
@bchesky
Follow
Marriott wants to add 30,000 rooms this year. We will add that in the next 2 weeks.
Reply Retweet Favorite More HootSuite
RETWEETS 286
FAVORITES 228
11:11 PM - 10 Jan 2014

TORONTO
YOUR uberX IS ARRIVING NOW

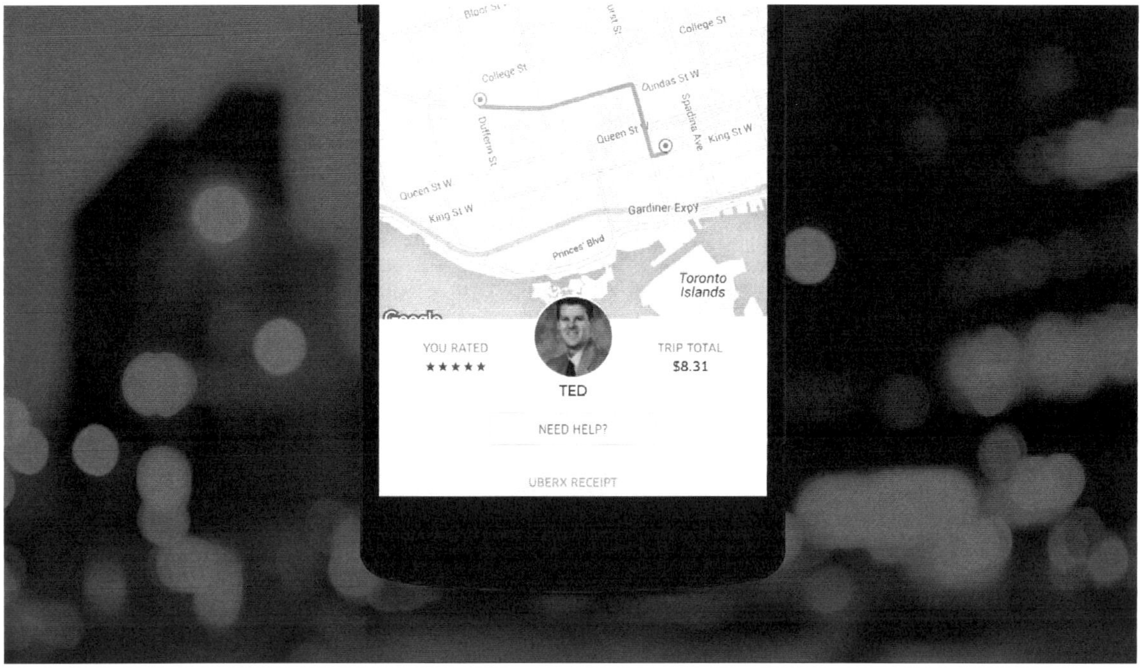
Bloor St
College St
College St
Dundas St W
Dufferin St
Spadina Ave
Queen St W
King St W
Queen St W
King St W
Gardiner Expy
Princes' Blvd
Toronto
Islands
YOU RATED
★★★★★
TED
TRIP TOTAL
$8.31
NEED HELP?
UBERX RECEIPT

1
PARTNERS

2
FEEDBACK

Textile Museum
Canada
Bell Trinity
Square
Nathan Phillips
Square
483 Bay St, Toronto, ON M5G 1P5,
Canada
4.6 ★ | UBERX

3
RISK &
REWARD

Home
Notifications
Messages
Search Twitter
Tweet
What's happening?
TO UberX Mugshots retweeted
N @Nakish19_77 May 4
@InsuranceBureau @OttawaBylaw @SJones_EPS_OTT this car is a gypsy diving passenger around, please stop him!
3
1

Profile
the uber of
the uber of
The Uber Of
@TheUberOf
Ted Graham
Innovation Leader at PwC
Follow
How many companies are claiming to be the "Uber of" their industry?
Feb 12, 2015 3,703 views 216 Likes 22 Comments

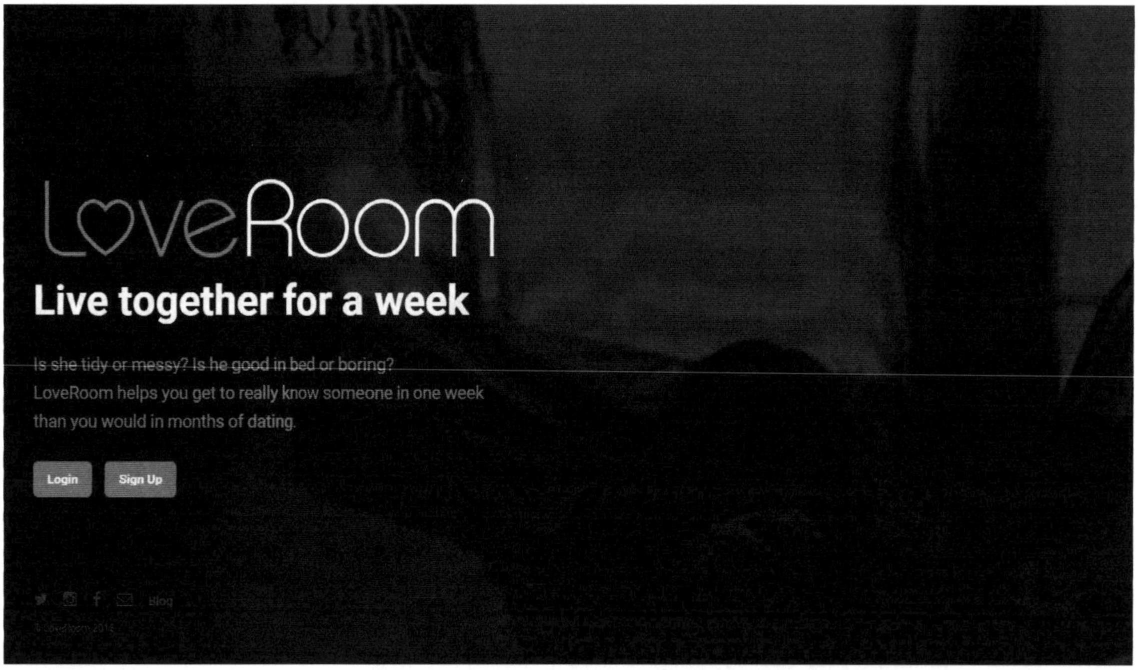
LoveRoom
Live together for a week
Is she tidy or messy? Is he good in bed or boring?
LoveRoom helps you get to really know someone in one week than you would in months of dating.
Login
Sign Up

THE UBER OF…

BlackJet
PRIVATE JETS

BEAUTIFIED
BEAUTY

plowme
SNOW PLOWING

HOMEJOY
HOME CLEANING

DRIZLY
BEER DELIVERY

swifto
When you need it now
DOG WALKING

seamless
FOOD DELIVERY

shyp
SHIPPING

BY 2020 MORE THAN 50% OF THE WORKFORCE WILL BE FREELANCE…

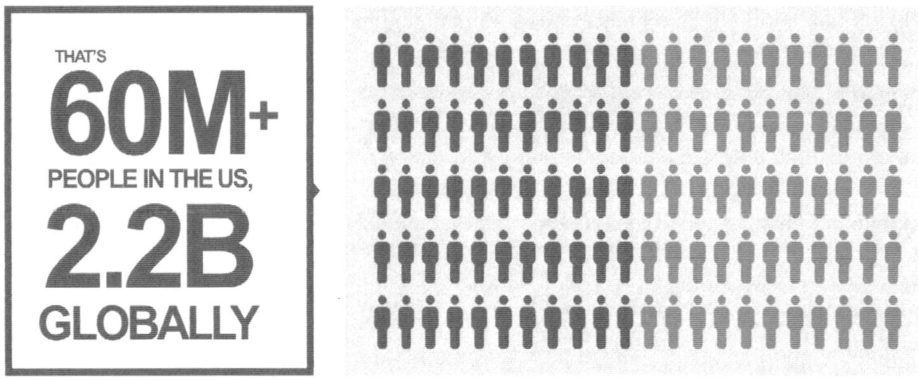

Intuit 2020 report

Average Annual USA Household Spending...

Housing = $17K

33% of total spend

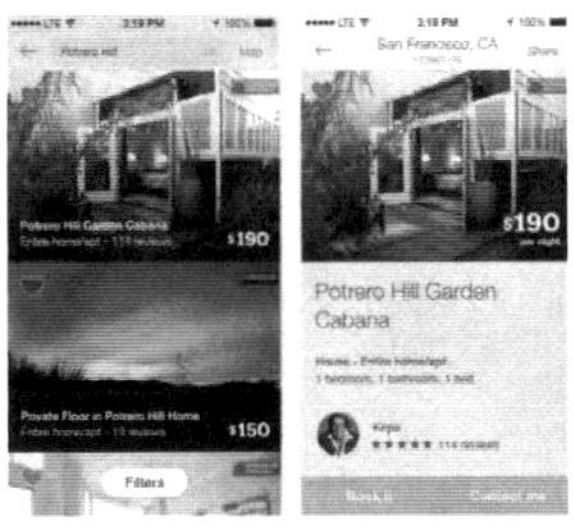

Transportation = $9K

18% of total spend

Food = $7K

14% of total spend

$4K on food consumed in home...
8% of spend

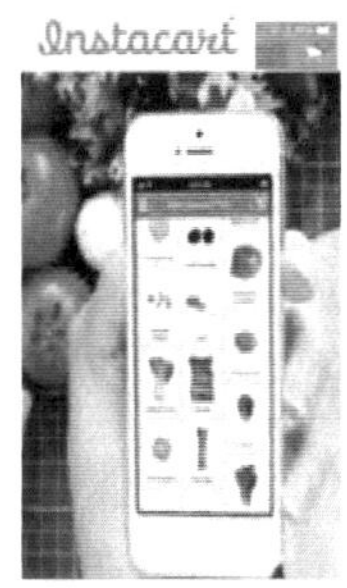

@KPCB Source: U.S. Bureau of Labor Statistics.
Image: Airbnb, Uber, Instacart.

A FREED MARKET ECONOMY?

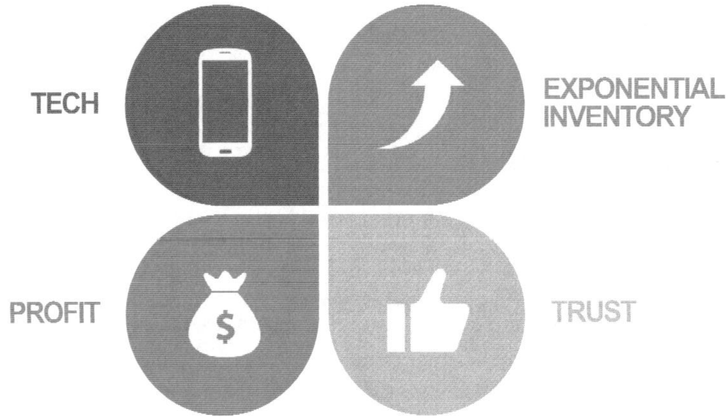

MOVING HUMANITY FORWARD
ZERO CRASHES,
ZERO EMISSIONS,
ZERO CONGESTION.
GENERAL MOTORS
CANADA

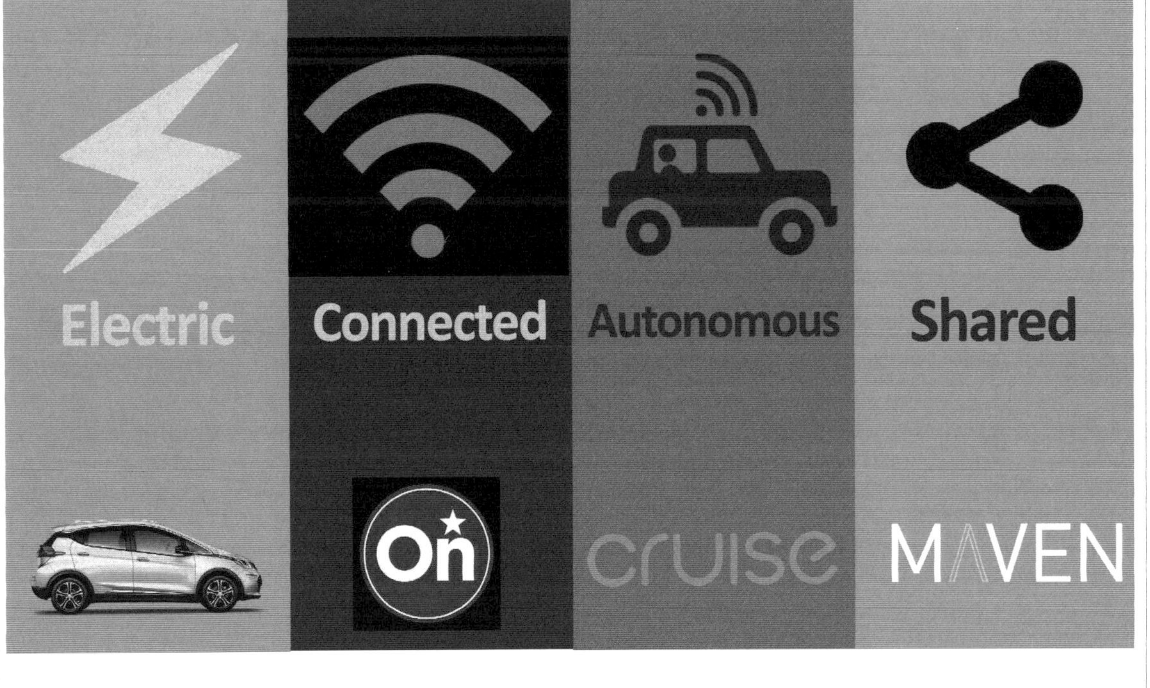
Electric
Connected
Autonomous
Shared
On
cruise
MAVEN

GENERAL MOTORS

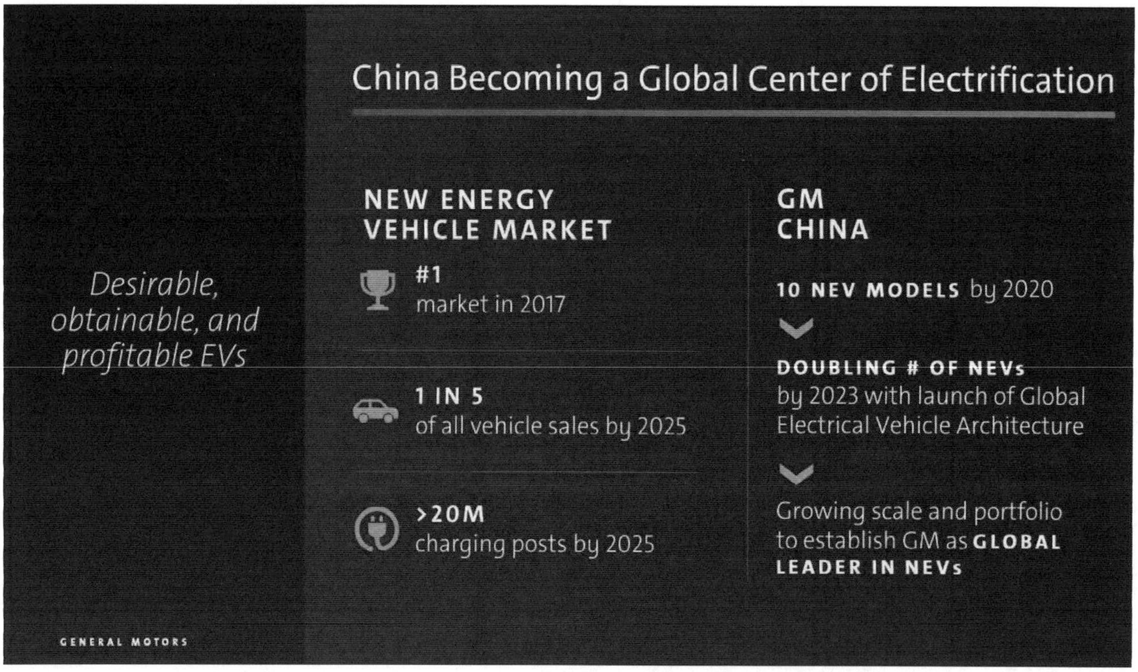
China Becoming a Global Center of Electrification
Desirable, obtainable, and profitable EVs
NEW ENERGY VEHICLE MARKET
#1
market in 2017
1 IN 5
of all vehicle sales by 2025
>20M
charging posts by 2025
GM CHINA
10 NEV MODELS by 2020
DOUBLING # OF NEVs
by 2023 with launch of Global Electrical Vehicle Architecture
Growing scale and portfolio to establish GM as GLOBAL LEADER IN NEVs
GENERAL MOTORS

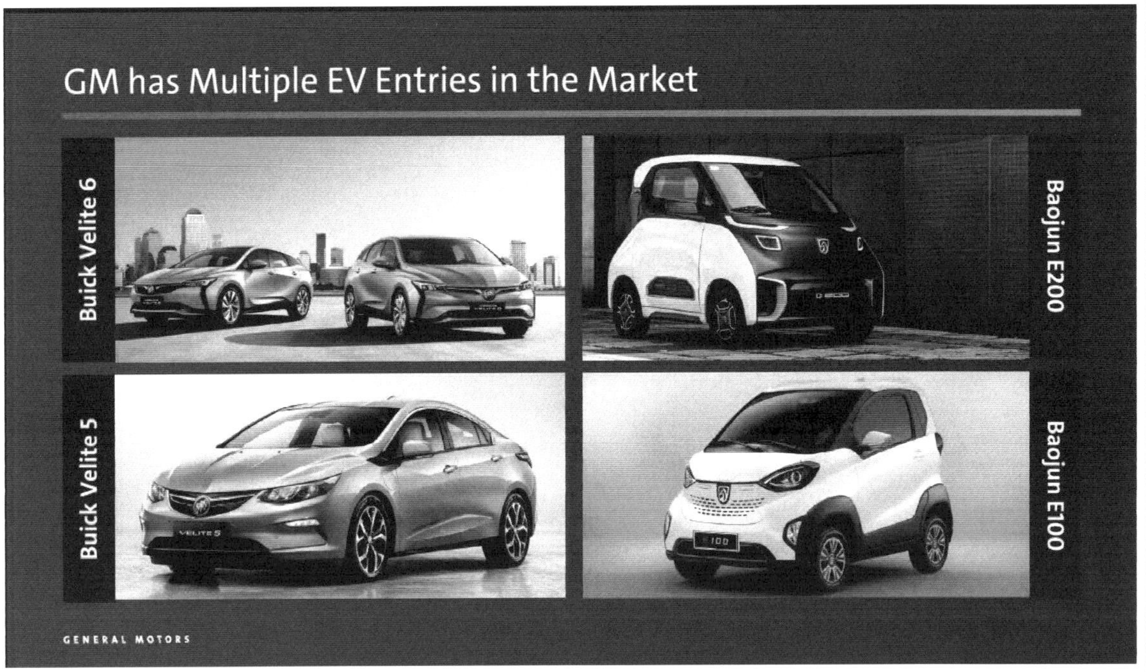
GM has Multiple EV Entries in the Market
Buick Velite 6
Baojun E200
Buick Velite 5
Baojun E100
GENERAL MOTORS

STATE OF THE ART EME 1.0 BATTERY ARCHITECTURE

BOLT BATTERY 〉 **ALL-NEW BATTERY SYSTEM** | **SIGNIFICANT IMPROVEMENTS**

Cell Cost
$145/kWhr

Cell Cost
$<$100/kWhr

- >30% cost reduction
- Larger footprint, lower height
- Higher energy density
- Flexible, modular design
- Improved DC Fast Charge

ALL-NEW EME 1.0 BATTERY PLATFORM PROVIDES FLEXIBLE PACK CONFIGURATIONS AT LOWER COST

GENERAL MOTORS 16

EV IS THE FOUNDATION FOR AUTONOMOUS VEHICLES

LEVERAGES GM ASSETS

SIMPLER INTEGRATION OF TECHNOLOGIES

OPTIMAL FOR URBAN ENVIRONMENT

SUPPORTS OUR ZERO EMISSIONS FUTURE WORLD VIEW

GENERAL MOTORS 18

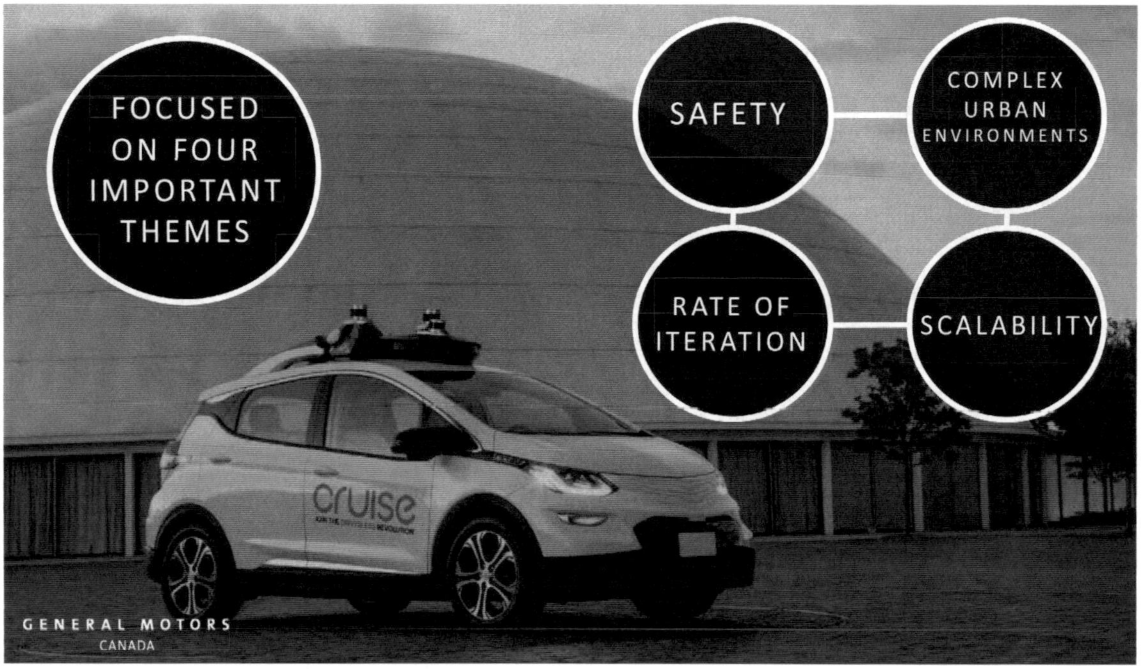
FOCUSED ON FOUR IMPORTANT THEMES
SAFETY
COMPLEX URBAN ENVIRONMENTS
RATE OF ITERATION
SCALABILITY
cruise
GENERAL MOTORS
CANADA

GENERAL MOTORS
CANADA
ZERO CRASHES

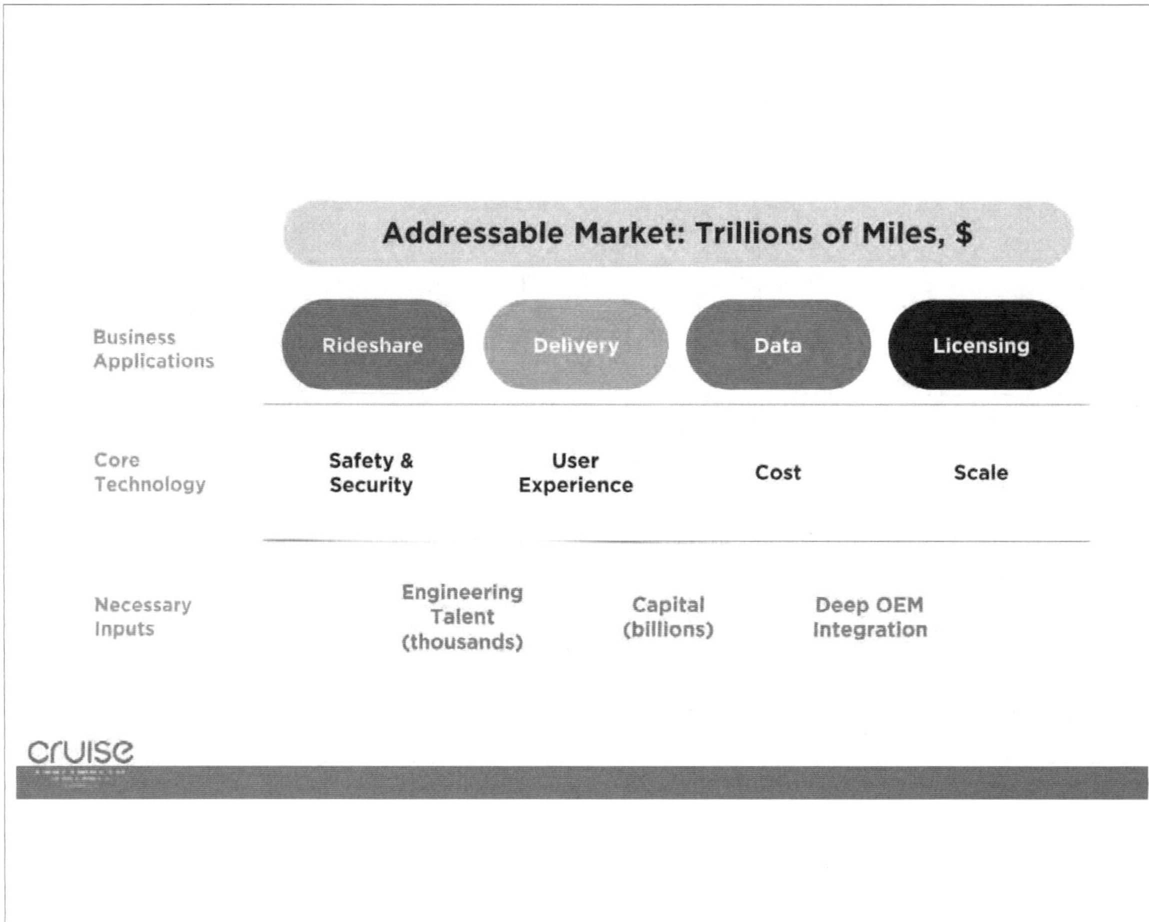

Addressable Market: Trillions of Miles, $
Business Applications
Rideshare
Delivery
Data
Licensing
Core Technology
Safety & Security
User Experience
Cost
Scale
Necessary Inputs
Engineering Talent (thousands)
Capital (billions)
Deep OEM Integration
cruise

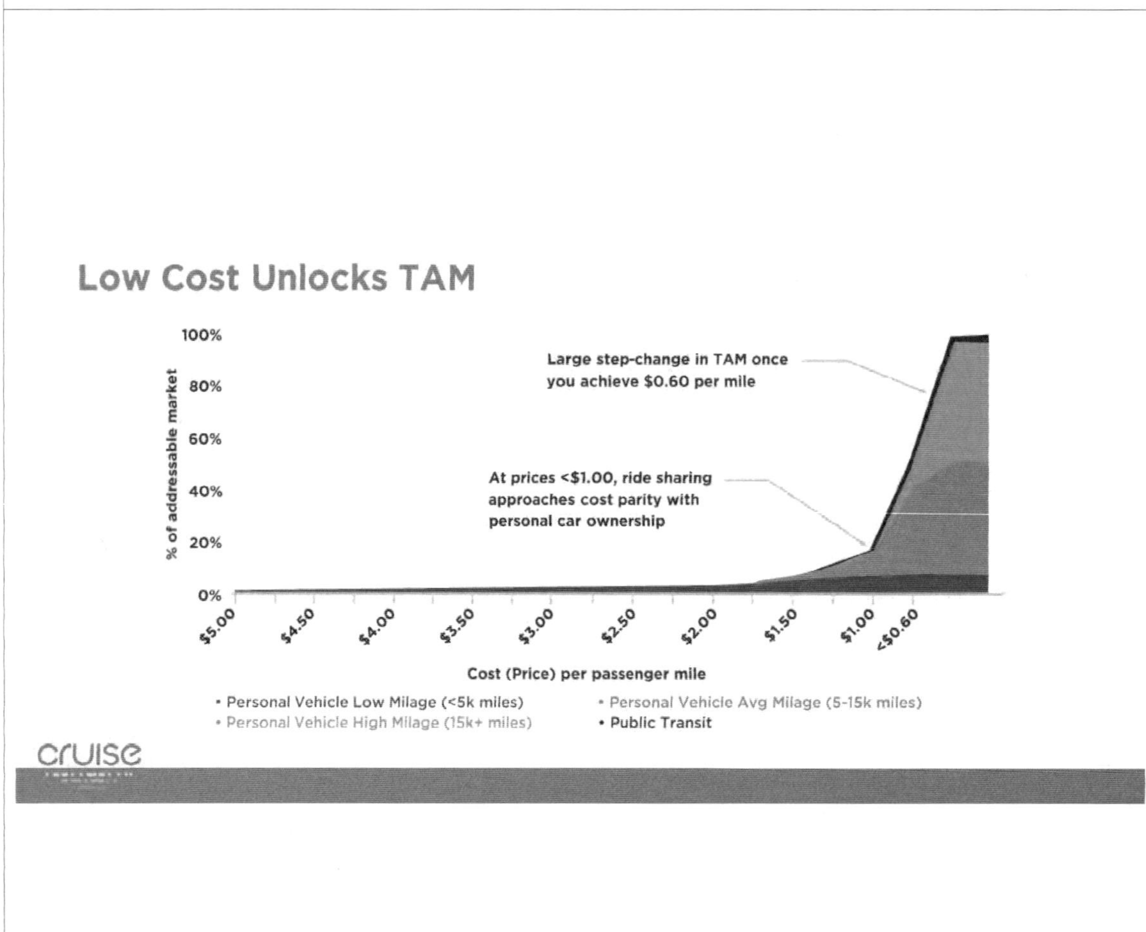

Low Cost Unlocks TAM
100%
80%
60%
40%
20%
0%
% of addressable market
Large step-change in TAM once you achieve $0.60 per mile
At prices <$1.00, ride sharing approaches cost parity with personal car ownership
$5.00
$4.50
$4.00
$3.50
$3.00
$2.50
$2.00
$1.50
$1.00
<$0.60
Cost (Price) per passenger mile
• Personal Vehicle Low Milage (<5k miles)
• Personal Vehicle Avg Milage (5-15k miles)
• Personal Vehicle High Milage (15k+ miles)
• Public Transit
cruise

GENERAL MOTORS
CANADA
3) ZERO CONGESTION

M/\VEN
CAR SHARING
Where do you want to drive?
Canada
Next
Back
M/\VEN
Toronto
Find A Car
My Trips
Help
More
M/\VEN
TRIP START
TRIP END
NOW
NOT-FOR-RENT
2016 Chevrolet Tahoe - Red
2016 Chevrolet Malibu - Red
GENERAL MOTORS
CANADA
ZERO CONGESTION

AREAS OF NEED & OPPORTUNITY
Lightweight Materials
Mobile Connectivity
Data Analytics
Advanced Battery Technology
Cyber Security
Software Development
Sensors
Artificial Intelligence

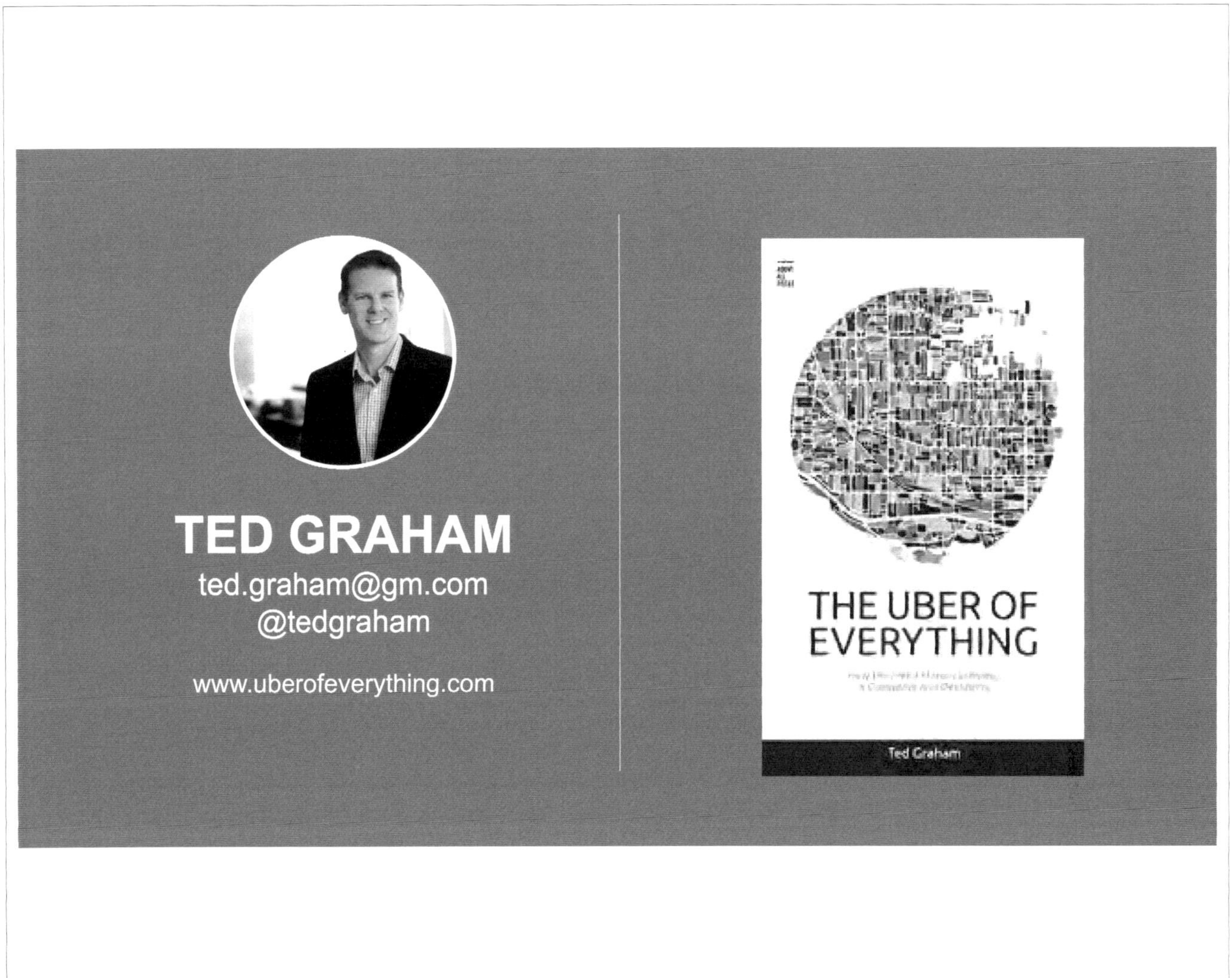
TED GRAHAM
ted.graham@gm.com
@tedgraham
www.uberofeverything.com
THE UBER OF EVERYTHING
Ted Graham

Regulatory Update: Challenges to the Market

Eric Van Genderen, Ph.D.
Associate Director, Environment & Sustainability
International Zinc Association

Overview

Chemical Management

- North America – Canada CMP3, USEPA, California
- Latin America – GHS
- Europe – REACH
- Asia-Pacific – K-REACH, Aust/NZ, ROA
- Global – IMO/MARPOL, Minamata

Zinc Product Stewardship

Consequences of Dis-Harmonization

- Market access challenges associated with mosaic of classifications
- Migration of inappropriate GHS classifications into developing economies
- Classifications for substances in consumer products are prohibitive

<u>Required</u> GHS Symbol:
Reproductive Toxicity (Cat. 2)
STOT-SE & STOT-RE (Cat. 1 & 2)

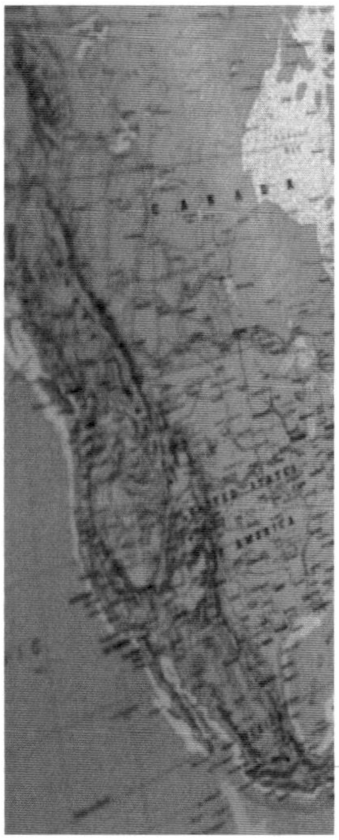

North America

Canada

- Zn WQG updated Sept, 2018
- Draft CMP3 Zn Assessment due Spring 2019

USA

- Zn AWQC/Stormwater updates coming

California

- DTSC "Zinc in Tires" petition ongoing
- Prop 65 Safe Use Designation for <u>all</u> metallic zinc applications

International Zinc Association

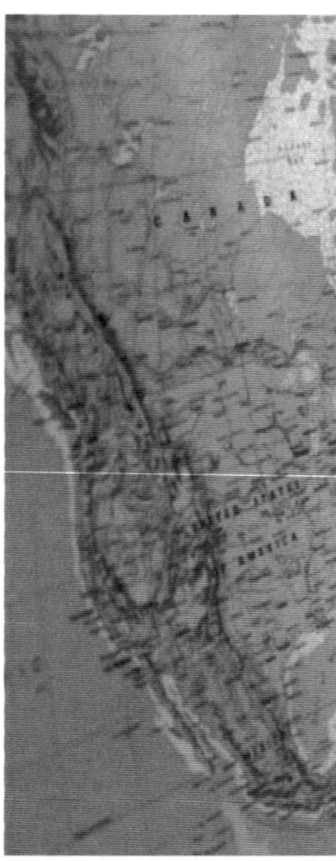

Canada

- NEW Water Quality Guideline (Sept, 2018)
  - ➤ 30-fold range; relative to bioavailability
- CMP3 "Zinc Group" Assessment
  - ➤ ECCC/HC announced Spring 2019 public release of draft report
  - ➤ Zinc Coalition fly in with Ottawa authorities
  - ➤ **Need industry representation to protect markets!**

International Zinc Association

CMP Risk Mgmt Potential

- Schedule 1 addition (TSL)
- Significant new activity notices
- Pollution prevention planning
- Water/Soil/Sediment guidelines for environment and health
- Waste-water regulations
- Phase out certain consumer products
- Emergency regulation requirements (Vanadium)
- Code of Practice for metal mining (Cobalt)
- Product reformulation (Boron)
- Targeted use reductions (Boron)
- Environmental Code of Practice for non-Integrated steel mills
- Substance-specific regulations

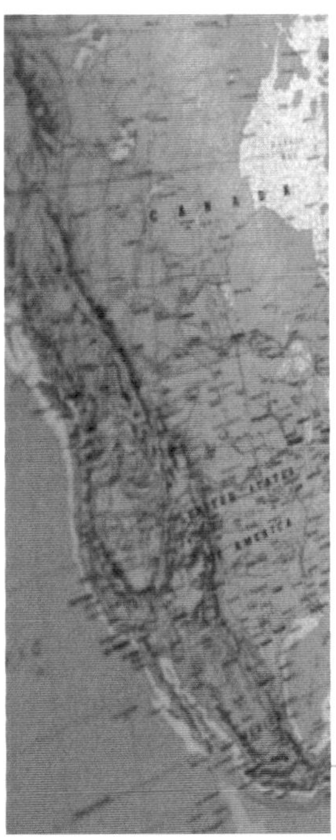

United States of America – USEPA –

- Promoting bioavailability updates since 2006
- Formal agreement (CRADA) with 8 associations
  - ➢ Sweeping guidelines for metals (2020)
  - ➢ Updates for Cu, Ni, Pb, and Zn (2023)
- 2020 Draft MSGP (Stormwater) update
  - ➢ National Academies Report highlights bioavailability

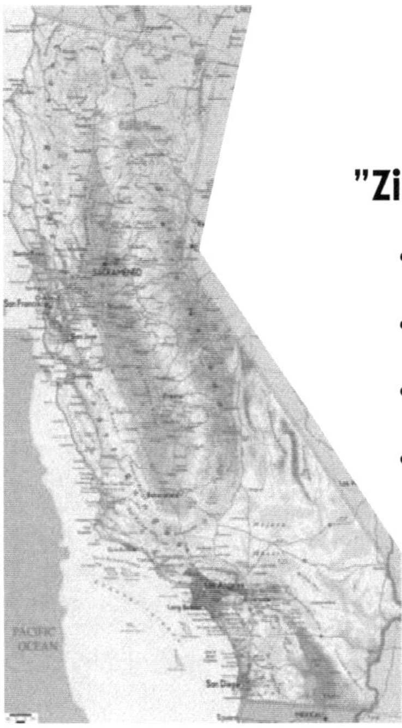

California – DTSC

"Zinc in Tires"

- May 2018 – CASQA filed petition
- July 2018 – IZA/USTMA arguments
- Jan 2019 – DTSC Supplemental Info Request
- Mar/Apr 2019 – IZA/USTMA advocacy

U.S. TIRE MANUFACTURERS ASSOCIATION CASQA CALIFORNIA STORMWATER QUALITY ASSOCIATION Department of Toxic Substances Control

International Zinc Association

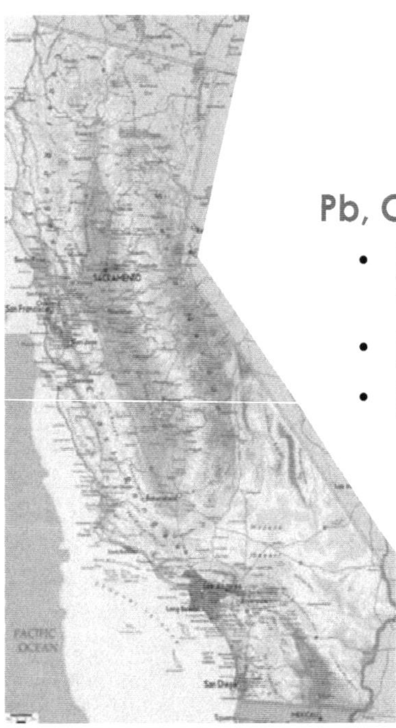

California – Prop 65

Pb, Cd, Ni exposures in consumer products

- Now: Metallic zinc impurities targeted as basis for financial penalties
- Next: "Position Paper" available Summer 2019
- Eventually: CA "Safe Use Designation" petition

Application	Maximum Pb (% bt wt)	Maximum Cd (% by wt)
Galvanizing	0.003 – 1.4	0.003 – 0.2
Die Casting	0.003 – 0.006	0.002 – 0.006
Rolled Zinc	<0.005	<0.005

International Zinc Association

Latin America

MEXICO
- Draft Hazardous Chemicals or Mixtures In Workplaces
- Mandatory compliance by Oct 2018

COLOMBIA
- Most progressive
- GHS reaches into consumer products
- Update of labels and SDSs every three years

CHILE
- Draft Regulation on Classification, Labeling, and Notification (Nov 2017)

COSTA RICA
- 5 year transition period to comply with labeling
- Includes consumer goods

BRAZIL

RoHS

CONASQ:
- registry of chemical production and imports
- risk assessment process
- risk management program

ANTT:
- Transport on highways and railways (Dec 2018)

ARGENTINA
- Draft hazardous waste regulation
- Access to potable water and sanitation are "essential human rights."

EU REACH

ZINC REACH	Cd REACH	Indium REACH	REACH Ge Germanium
• Zinc metal • 13 substances (e.g., ZnO, nano-ZnO, salts) • 7 Intermediates (e.g. calcines)	• Cadmium metal • 11 substances (e.g., CdO, pigments, CdTe)	• Indium metal • 7 substances (e.g., In2O3, InCl3)	• Germanium metal • GeO2, GeCl4

1. Registration complete for all substances >10 T/a
2. Optimize files for Evaluation (avoid risk management)
3. Focus on cadmium Authorization

Korea REACH

A. Inappropriate registration requirements
No clear definition of Naturally Occurring Substance and Waste, absence of recognition of relevant waivers and adaptations.

B. Issues with using EU dossier content in Korea
No evidence of applying read-across approach; limited application of EU registration dossiers.

C. Strict regulation with limited flexibility
Disproportionate burden of proof on industry; not based on applicable scientific concepts.

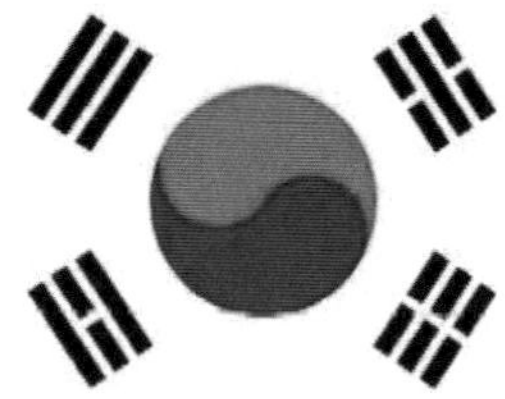

Toxic Chemical Control Act

International Zinc Association
13

Australia/New Zealand

NICNAS and HSNO

- Harmonized classification and assessment with EU REACH coordinated through MCA
- AUS: July 2020 commencement (7th GHS)

CSIRO and NIWA

- 2016 draft Water Quality Guideline Value prohibitively low
- IZA engaged with authorities

International Zinc Association

Other Asia

China
- Draft ERA Guidance (Jan 2019)
- Risk Mgmt guidance
- Native language limitation

Thailand
- TECI (hazard-based)
- 2019 release?
- Authoritative prioritization

Vietnam
- NCI
- Priority nominations (2018/19)
- ASEAN database

Philippines
- Cadmium in focus?

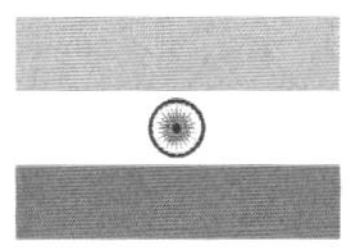

India
- National Action Plan

International Zinc Association

MARPOL Annex V:
Harmful to the Marine Environment (HME)

- HME declaration now mandatory part of the shipper's declaration
- Some confusion by owners over the status
  - ➢ Self-classification
  - ➢ Confidence in declaration

International Zinc Association

Definition of Group A Cargoes

- Group A cargoes proposed re-definition:
  - "*Group A* consists of cargoes which due to their moisture content may adversely affect the stability [and/or safety] of the ship, and[/or] may liquefy if shipped at a moisture content in excess of their transportable moisture limit."
- Member States and international organizations open comments on amendments:
  - "Group A consists of cargoes, which due to their moisture content, may adversely affect the stability of the ship, including the potential to liquefy, if shipped at a moisture content in excess of their transportable moisture limit."

International Zinc Association

IMSBC Code Amendment 05-19

- New consolidated version of IMSBC Code -refer E&T 30/WP.1 - Annex 1 (amendments to existing schedules) and Annex 3 (the complete new text)
- New individual schedule for Metal Sulfidic Concentrates, Self-heating, UN 3190 as a Group A and B cargo.
- New, separate schedules for "Flue Dust" and "Flue Dust, Zinc Refining" -referred to E&T 30.
  - E&T 30 modified "Flue Dust" to "FLUE DUST, CONTAINING LEAD AND ZINC"
  - E&T 30 modified "Flue Dust, Zinc Refining" to "ZINC OXIDE ENRICHED FLUE DUST"
  - Both schedules are included in Annex 1 to E&T 30/WP.1
- Copper and Lead, containing matte -referred to E&T 30
  - E&T modified to "MATTE CONTAINING COPPER AND LEAD"
  - Schedule included in Annex 1 to E&T 30/WP.1

International Zinc Association

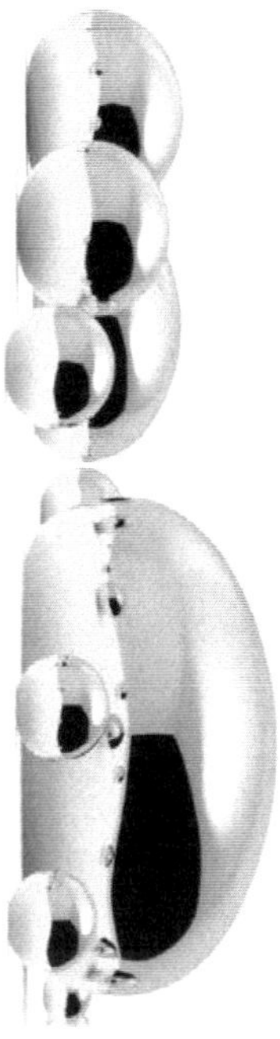

Mercury Waste

- UN regulation on Hg: Minamata Convention
  - Translated into national legislation
- What is Hg waste?
  - Relevant for classification of mining waste (overburden and tailings)
  - Based on hazard (content) or risk (solubility/exposure)?
- IZA participates via ICMM Working Group
- IZA Actions
  - Internal data collection (loads, BAT, etc.)
  - Develop position (natural, solubility, assessment methods)

IZA and its members are committed to the principle of sustainability, and this commitment is embedded in our <u>sustainability charter</u> and <u>guiding principles</u>. We believe that protection of the environment, open engagement on sustainability issues, and supporting sustainable development practices not only drive long-term prosperity for the zinc industry; they enable our customers to become more sustainable through the use of zinc products.

SUSTAINABLE DEVELOPMENT GOALS

IZA Focus on UN SDGs

https://sustainability.zinc.org/

Desk Strategy |

MACQUARIE

Zinc Market Outlook

Not going down without a fight

Vivienne Lloyd
vivienne.lloyd@macquarie.com
Macquarie Commodities Strategy
Macquarie Capital (Europe) Ltd
28 Ropemaker Street
London EC2Y 9HD, UK

February 2019

Desk Strategy |

MACQUARIE

Zinc this time last year was peaking – we called for prices to fall and they did….

LME zinc cash price ($/t)

Source: LME, Macquarie Commodities Strategy, February 2019

Desk Strategy |

MACQUARIE

2018: The new mines came along, the TCs soared, the price faded – and then the correlation broke

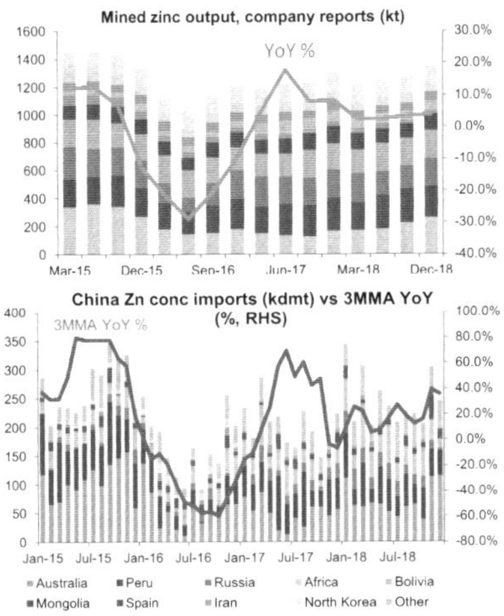

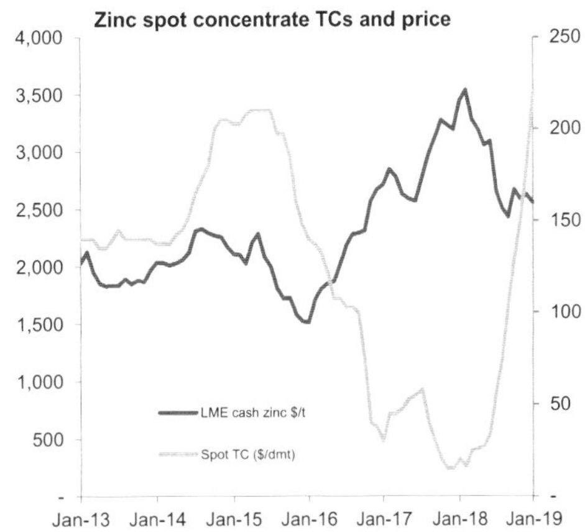

Source: Company reports, China Customs, Wood Mac, LME, Macquarie Commodities Strategy, February 2019

PAGE 3

Desk Strategy |

MACQUARIE

Key catalysts in H2 2018

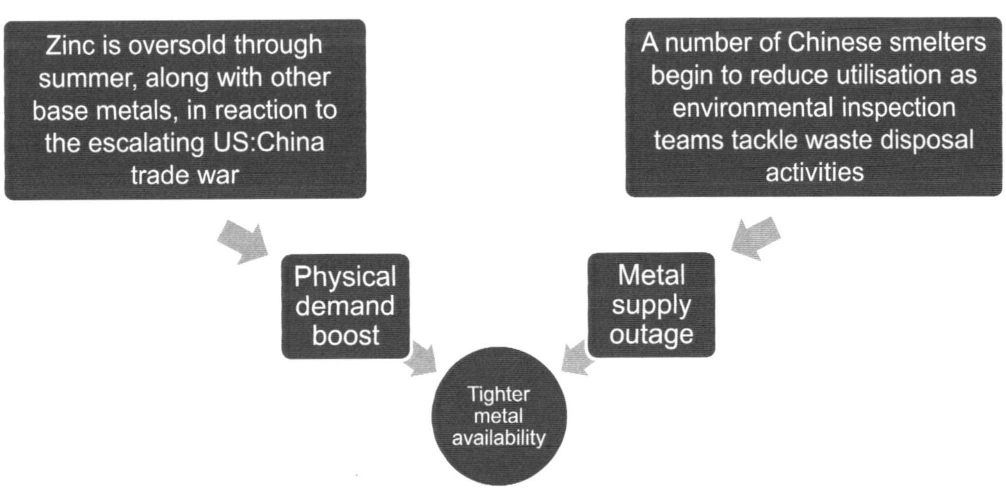

PAGE 4

Desk Strategy |

MACQUARIE

Sounds good – except the arb didn't really back the story, it swung shut in late 2018 as too much ingot showed up

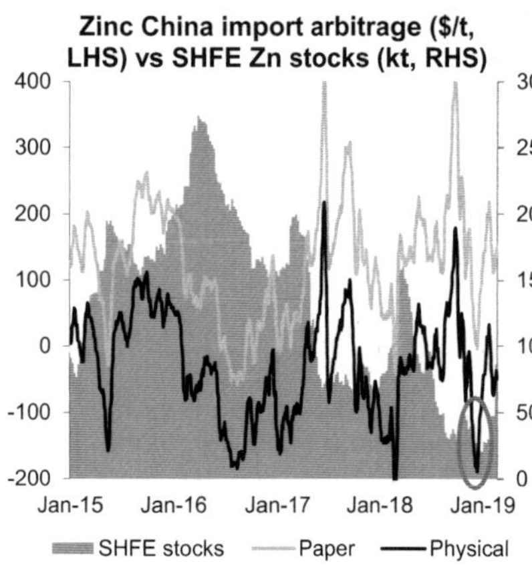

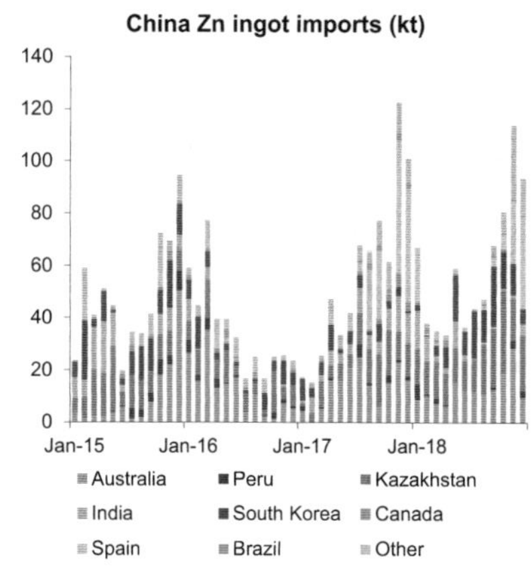

Source: Company reports, China Customs, Wood Mac, LME, Macquarie Commodities Strategy, February 2019

Desk Strategy |

MACQUARIE

Demand: global growth a little less commodity-friendly

- Industrial production, more commodity-intensive than GDP as a whole, weakened more than GDP through much of 2018, and while it has bounced in the last few months it is not very convincing.
- In terms of the PMI Europe main driver of the sharp fall seen over the past year. There autos have been a factor behind the slump, and car production is recovering, though it has not turned around wider IP yet. Some calendar factors at work too..

Industry runs out of steam

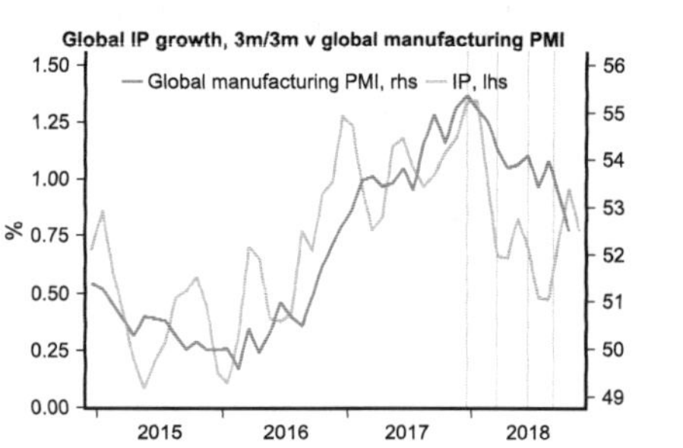

Source: National Sources, Macquarie Commodities Strategy, February 2019

Eurozone a major contributor

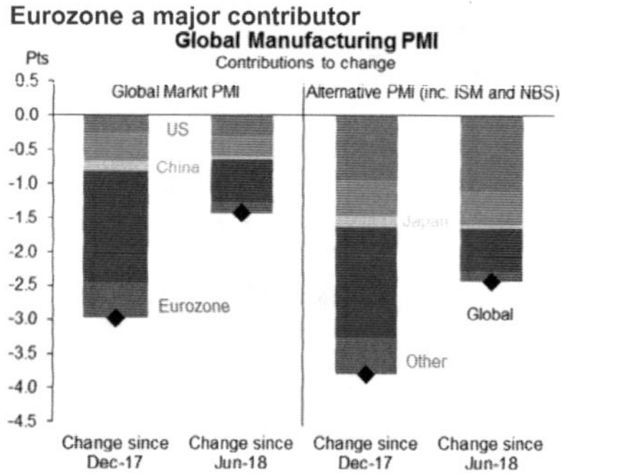

Source: Macrobond, Macquarie Commodities Strategy, February 2019

Desk Strategy |

MACQUARIE

Sinking autos sales hurts zinc (and Pb) the most of base with both galv steel and diecast alloys hit

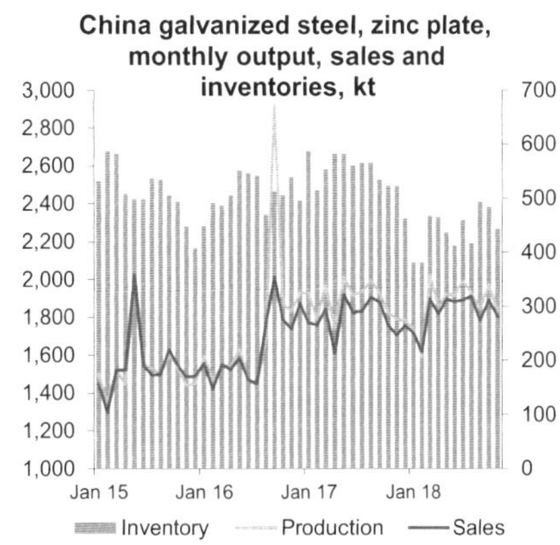

Source: National data sources, Wood Mac, LME, China Customs, Macquarie Commodities Strategy, February 2019

Desk Strategy |

MACQUARIE

Moreover, zinc's relative price trajectory since 2016 has undermined demand, leading to the lowest growth in base..

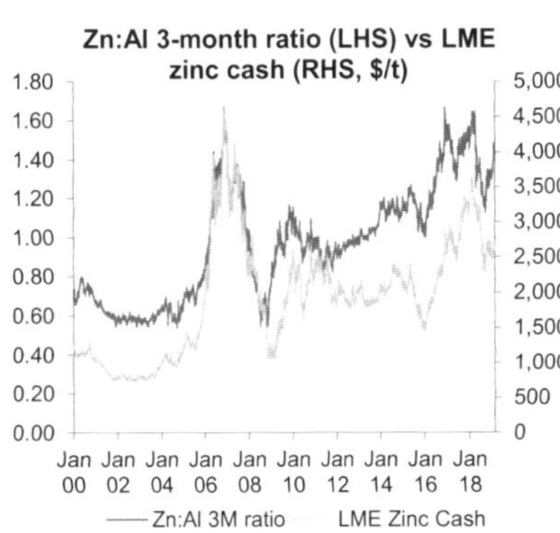

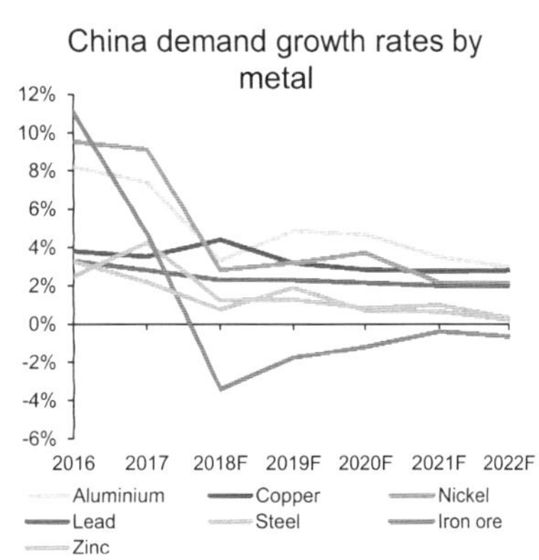

Source: LME, Wood Mac, LME, China Customs, Macquarie Commodities Strategy, February 2019

Desk Strategy |

MACQUARIE

And yet… the stock draws have been huge, the LME is emptying, and the forward curve is highly volatile

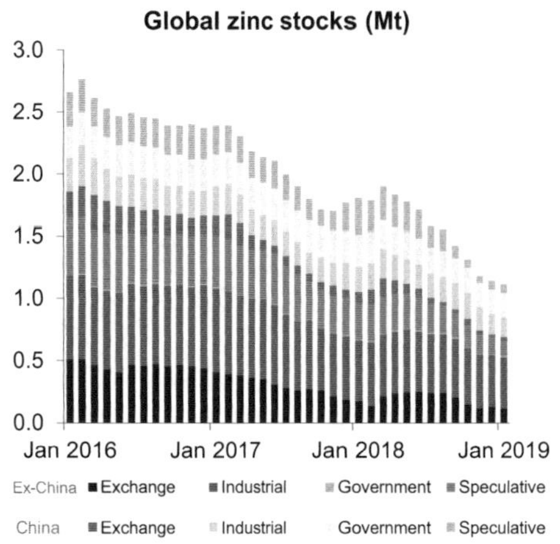

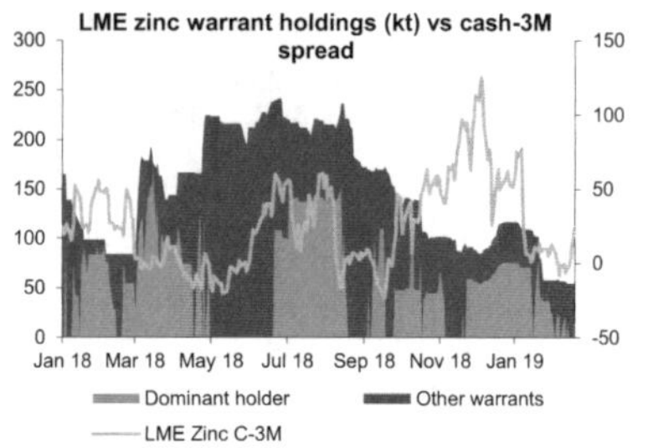

Source: Company reports, China Customs, Wood Mac, LME, Macquarie Commodities Strategy, February 2019

PAGE 9

Desk Strategy |

MACQUARIE

Chinese refined output problems creating a bottleneck, keeping metal tight while driving up the TCs

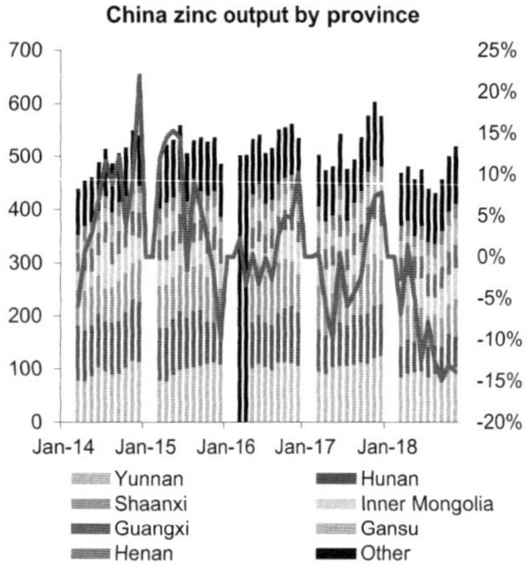

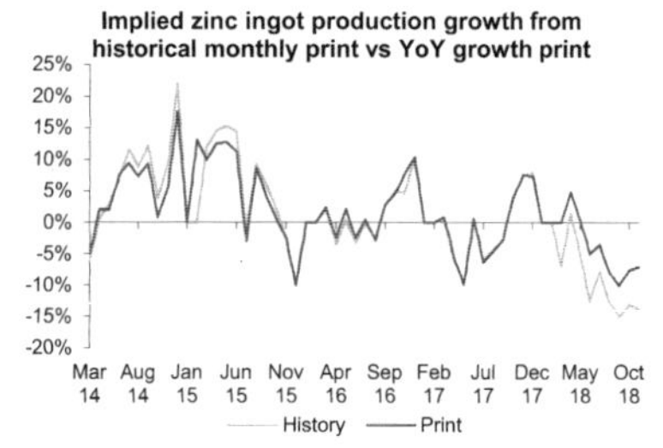

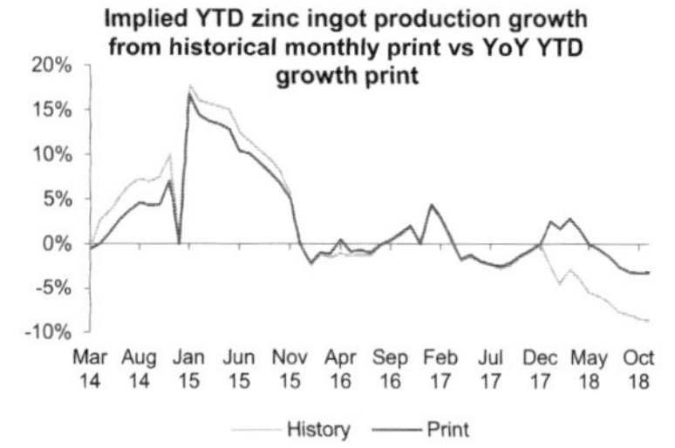

Source: CEIC, NBS, Macquarie Commodities Strategy, February 2019

PAGE 10

Desk Strategy |

MACQUARIE

Also key – copper is rallying on supply issues plus the easing of trade conflict hostilities

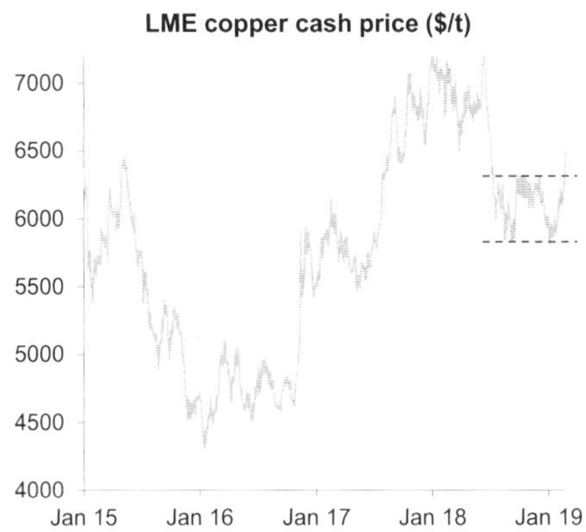

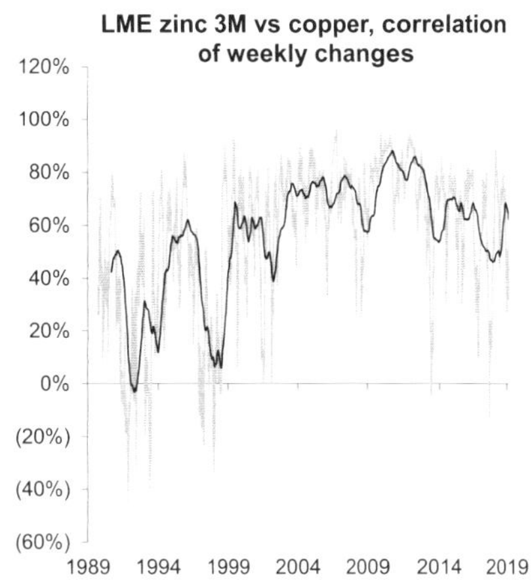

Source: LME, Macquarie Commodities Strategy, February 2019

Desk Strategy |

MACQUARIE

Thus, we acknowledge resistance to the bear narrative; Zn will be lower in 6mo and beyond, but nearby likely to move higher

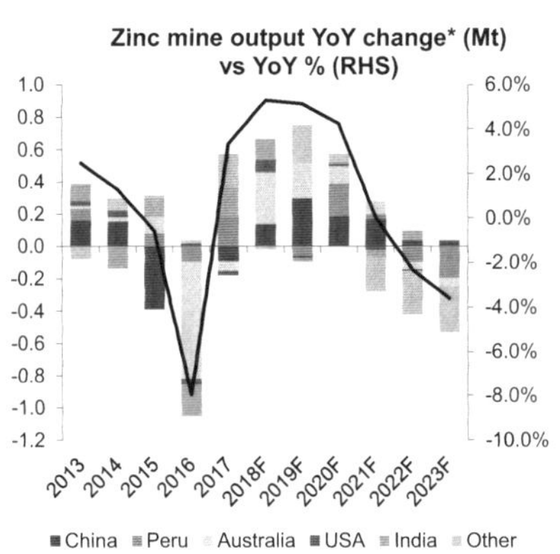

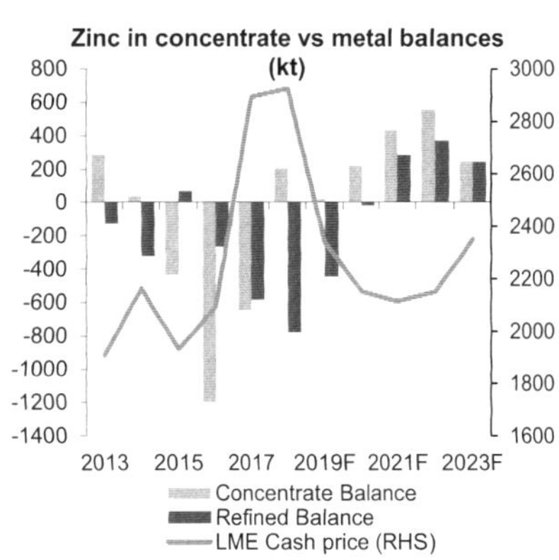

*before disruption

Source: Wood Mackenzie, BGRIMM, LME, ILZSG, Metal Bulletin, Macquarie Commodities Strategy, February 2019

Desk Strategy |

MACQUARIE

Supply-Demand balance – Zinc

'000t zinc	2015	2016	2017	2018	2019F	2020F	2021F	2022F	2023F
Mine production	12,766	11,749	12,140	12,784	13,032	13,811	14,488	14,842	14,575
% Change YoY	-0.6%	-8.0%	3.3%	5.3%	1.9%	6.0%	4.9%	2.4%	-1.8%
Concs Balance	**-438**	**-1199**	**-648**	**201**	**18**	**221**	**434**	**555**	**244**
Refined production	13,700	13,557	13,489	13,354	13,889	14,514	15,022	15,322	15,475
% Change YoY	3.4%	-1.0%	-0.5%	-1.0%	4.0%	4.5%	3.5%	2.0%	1.0%
Consumption	13,633	13,825	14,072	14,130	14,336	14,531	14,739	14,953	15,233
% Change YoY	0.5%	1.4%	1.8%	0.4%	1.5%	1.4%	1.4%	1.5%	1.9%
Refined Balance	**67**	**-268**	**-583**	**-775**	**-447**	**-17**	**283**	**369**	**242**
Estimated total stocks	2723	2369	1771	1142	694	677	960	1328	1572
Weeks of consumption	10.4	8.9	6.5	4.2	2.5	2.4	3.4	4.6	5.4
LME Cash Price ($/t)	**1,932**	**2,092**	**2,894**	**2,925**	**2,338**	**2,150**	**2,113**	**2,150**	**2,350**
LME Cash Price (c/lb)	88	95	131	133	106	98	96	98	107

Source: ILZSG, Wood Mackenzie, LME, Macquarie Commodities Strategy, February 2019

Desk Strategy |

MACQUARIE

AUTHOR INDEX

International Zinc Association
2530 Meridian Parkway
Suite 115
Durham, NC 27713
USA

ISBN 978-1-5108-8559-2

9 781510 885592

CURRAN ASSOCIATES INC.
proceedings.com